Recursive Analysis

R. L. Goodstein

Dover Publications, Inc.
Mineola, New York

D0023823

Bibliographical Note

This Dover edition, first published in 2010, is an unabridged republication of the work originally published in 1961 by North-Holland Publishing Company, Amsterdam.

Library of Congress Cataloging-in-Publication Data

Goodstein, R. L. (Reuben Louis)
Recursive analysis / R. L. Goodstein. — Dover ed.
p. cm.
Originally published: Amsterdam : North-Holland Pub., 1961.
Includes bibliographical references and index.
ISBN-13: 978-0-486-47751-0
ISBN-10: 0-486-47751-7
1. Recursion theory. I. Title.

QA9.6.G66 2010
511.3'5—dc22

2009043065

www.doverpublications.com

PREFACE

There are several systems of constructive function theory; some, like intuitionist analysis are based on a new logic, some study constructive objects, recursive real numbers for example, by the methods of classical analysis, and some seek to translate a portion of classical analysis into free variable formulae by means of functionals. The system described in this book differs from all of these. It is based on primitive recursive arithmetic and seeks to build up a function theory of the rational field which in some respects closely resembles classical analysis and in others is closer to intuitionist analysis. All proofs in recursive analysis are formalisable in the equation calculus, the system of recursive arithmetic described in my book *Recursive Number Theory*, but the present work may be read without a detailed knowledge of recursive arithmetic.

I am again deeply indebted to Professor Heyting for his very kind encouragement, to Mr. John Hooley for generous help in reading the proofs, and to the compositors and printers of the North Holland Publishing Company for the pleasure they have given me by the excellence of their work.

R. L. Goodstein

The University, Leicester, England

SYMBOLS

CONTENTS

CHAPTER I

RECURSIVE CONVERGENCE

Primitive and general recursive functions. Recursive arithmetic and its extensions. Recursive convergence and relative convergence. The reduced sequence. Recursive limits and tests for recursive convergence. Primitive and general recursive real numbers.

1 Recursive analysis is a free variable theory of functions in a rational field, founded on recursive arithmetic. It involves no logical presuppositions and proceeds from definition to theorem by means of derivation schemata alone.

The elementary formulae of recursive arithmetic are equations between *terms*, and the class of formulae is constructed from the elementary formulae by the operations of the propositional calculus. The terms are the free numeral variables, the sign 0 and the signs for functions. The function signs include the sign $S(x)$ for the successor function (so that $S(x)$ plays the part of $x+1$ in elementary algebra) and signs for functions introduced by recursion. The derivation rules are taken to be sufficient to establish the universally valid sentences of the propositional calculus, and include a schema permitting the substitution of terms for variables, the schema for equality

$$a=b \rightarrow \{\mathscr{A}(a) \rightarrow \mathscr{A}(b)\},$$

and the induction schema

$$\frac{\mathscr{A}(0),\ \mathscr{A}(n) \rightarrow \mathscr{A}(S(n))}{\mathscr{A}(n)},$$

the schemata for explicit definition of functions for any number of arguments, and finally schemata for definition by recursion. The simplest definition schema for recursion, the schema of *primitive recursion*, is

$$f(0, a)=\alpha(a),\ f(S(n), a)=\beta(n, a, f(n, a)).$$

Specifically this schema defines $f(n, a)$ by primitive recursion from the functions α and β. We take as initial primitive recursive functions the successor function $S(x)$, the identity function $I(x)$, defined explicitly by the equation $I(x)=x$, and the zero function $Z(x)$ defined by $Z(x)=0$. A function is said to be *primitive recursive* if it is an initial function or is defined from primitive recursive functions by substitution or by primitive recursion.

1, 1 In this work we shall be principally concerned with primitive recursive functions. Without changing the character of the system we shall build, the class of functions could be enlarged to include for instance *multiply recursive functions* and even certain ordinal recursive functions (of ordinal not exceeding $\omega^{\omega^{\omega}}$ at the present state of knowledge). The system could not however be enlarged to admit the class of functions which has played so important a part in foundation researches, the class of *quasi*- (or general) recursive functions, without changing the character of the system entirely. A quasi-recursive function is defined by a system of equations (on the right hand sides of which we may suppose only numerals or primitive recursive functions appear) from which, by substitution, the value of the defined function may be derived for each assigned set of values of the arguments. The left-hand side of the equations may contain, however, in addition to the function being defined, auxiliary functions, about which we may have only incomplete information. Consider for instance the set of equations [1])

$$\pi(x, y) = \prod_{n<y} \chi(x, n)$$

$$\tau(Sz, 0, y) = y$$

$$f(x) = \tau(\pi(x, y),\ \pi(x, Sy), y),$$

where

$$\prod_{n<0} \chi(x, n) = 1,\ \prod_{n<Sy} \chi(x, n) = \{\prod_{n<y} \chi(x, n)\}\cdot\chi(x, y) \text{ and } \chi(x, n)$$

is a primitive recursive function for which (in some sense which remains to be considered) it is known that for each x there is a y for which $\chi(x, y)=0$. Then these equations define $f(x)$ as a quasi-

[1]) cf. S. C. Kleene [2] pp. 279–280.

recursive function. To show this we have to consider the way in which the values of $f(x)$ may be derived from the equations. For the purpose of illustration let us suppose that for some given value of x, say $x=3$, the first value of y for which $\chi(x, y)$ vanishes is $y=7$. Then, keeping $x=3$, we see that

$$\pi(3, 0),\ \pi(3, 1),\ \pi(3, 2),\ \ldots,\ \pi(3, 7)$$

are all greater than zero, but $\pi(3, 8)=0$. The auxiliary function $\tau(u, v, y)$ is defined only for values of u greater than zero and for $v=0$; thus $f(3)$ can be evaluated only when we find a y for which $\pi(x, y)>0$ and $\pi(x, Sy)=0$ and, as we have supposed, this occurs when $y=7$ (and not for any other value of y), showing that $f(3)=7$. This illustrates the fact that to determine the value of $f(x)$, we must first find the value of y for which $\chi(x, y)=0$.

In the definition of a quasi-recursive function no limitation is imposed upon the way in which we show that, for each x, there is a y for which $\chi(x, y)=0$; we may for instance establish the existence of y by a proof in some formal system that the assumption that $\chi(x, y)>0$ for all y leads to a contradiction. Such a proof may provide us with no means of finding the y in question and therefore no means of finding $f(x)$; we may go on calculating the values of $\chi(x, y)$ for greater and greater values of y for untold time without finding the y for which $\chi(x, y)=0$. By confining our considerations to primitive recursive functions we are assured that the values of all functions in the system may be determined in a preassignable number of steps.

The most appropriate system for utilising quasi-recursive functions would be, not a free variable system, but one which contained an existential operator "$\exists$", and a minimal operator "μ" which selected the least of a given set of values. From an existential theorem

$$(\exists y)\ R(x, y)$$

the minimal operator derives a function $f(x)$ such that

$$f(x) = (\mu y)\ R(x, y).$$

If $R(x, y)$ is a primitive recursive relation then $f(x)$, defined by the minimal operator may be shown to be quasi-recursive; in fact the system of equations considered above is an instance of this. For future reference we list some further important properties of quasi-recursive functions and relations. [1])

There is a primitive recursive function $T(z, x, y)$ such that for $z=0, 1, 2, \ldots, (\exists y)T(z, x, y)$ is an enumeration of *all* predicates of the form $(\exists y)R(x, y)$ with recursive R. It follows that the class of predicates of the form $(\exists y)R(x, y)$ is the same whether R is quasi-recursive, or restricted to primitive recursive relations.

For any given R let r be a value of z for which

$$(\exists y)R(x, y) \leftrightarrow (\exists y)T(r, x, y)$$

(where $\leftrightarrow$ denotes equivalence of relations) and therefore

$$(\exists y)R(r, y) \leftrightarrow (\exists y)T(r, r, y)$$

and therefore $(\exists y)R(r, y)$ is *not* equivalent to $(y) \sim T(r, r, y)$, and (taking $\sim S$ for R, and s for r)

$$(y)S(s, y) \leftrightarrow (y) \sim T(s, s, y)$$

so that

$$(y)S(s, y) \text{ is not equivalent to } (\exists y)T(s, s, y).$$

Given any quasi-recursive predicate $R(x)$ let $R(x, y)$ denote the predicate $R(x) \& y=y$ so that $R(x, y)$ is also quasi-recursive and

$$R(x) \leftrightarrow (\exists y)R(x, y) \leftrightarrow (y)R(x, y)$$

whence it follows that there are numbers r, s for which

$$R(r) \text{ differs from } (y) \sim T(r, r, y)$$

and

$$R(s) \text{ differs from } (\exists y)T(s, s, y)$$

and therefore, *neither* $(\exists y)T(x, x, y)$ *nor* $(y) \sim T(x, x, y)$ are *recursive*. In other words $(\exists y)T(x, x, y)$ is an example of a predicate of the form $(\exists y)R(x, y)$, with primitive recursive R, which is *not* quasi-recursive.

---

[1]) *Vide* Kleene [2] pp. 281–282, 288.

Another important enumeration theorem runs as follows:

For a certain primitive recursive function $U(t)$ all quasi-recursive functions (of one variable) may be enumerated in the form

$$U(\mu y\, T(n, x, y)), \qquad n=0, 1, 2, \ldots$$

and with the same U, and a corresponding $T(n, x_1, x_2, \ldots, x_k, y)$ all quasi-recursive functions of k variables may be enumerated in the form

$$U(\mu y\, T(n, x_1, x_2, \ldots, x_k, y)), \qquad n=0, 1, 2, \ldots$$

To return to primitive recursive functions we may remark that there are many definition schemata which though seemingly very different from that of primitive recursion are known to define only primitive recursive functions. We shall not enter into a consideration of these schemata here but we shall have occasion in the sequel to note instances of such schemata. For an account of the forms of recursive definition reference may be made to the author's 'Recursive Number Theory' where it is shown that recursive arithmetic may be formalised in a system in which the axioms of the propositional calculus and the derivation schemata referred to above are all demonstrable.

1, 2 The development of recursive analysis from recursive arithmetic involves the introduction of rational numbers and functions. A rational number may be defined to be an ordered triplet $(p, q)/r$, of natural numbers, with $r>0$.

We define

$$(p, q)/r \gtreqless (p', q')/r' \leftrightarrow pr'+q'r \gtreqless p'r+qr'$$

where '$\leftrightarrow$' is the sign for equivalence; thus for instance the equation $(p, q)/r=(p', q')/r'$ holds if, and only if, $pr'+q'r=p'r+qr'$, and in particular $(p, q)/r=(kp+n, kq+n)/kr$, where $k>0$.

A recursive function of one (or more) variables $(p, q)/r$ is a triplet of recursive functions of natural numbers

$$\{P(p, q, r),\ Q(p, q, r)\}/R(p, q, r)$$

with $R(p, q, r) > 0$, and such that (writing P for $P(p, q, r)$, P' for $P(p', q', r')$ and so on)

$$(p', q')/r' = (p, q)/r \to (P', Q')/R' = (P, Q)/R.$$

For example the sum of two rational numbers is defined by

$$(p_1, q_1)/r_1 + (p_2, q_2)/r_2 = (p_1r_2 + p_2r_1, q_1r_2 + q_2r_1)/r_1r_2;$$

if

$$(p_1, q_1)/r_1 = (p_1', q_1')/r_1' \text{ and } (p_2, q_2)/r_2 = (p_2', q_2')/r_2'$$

then it is readily verified that

$$(p_1r_2 + p_2r_1, q_1r_2 + q_2r_1)/r_1r_2 = (p_1'r_2' + p_2'r_1', q_1'r_2' + q_2'r_1')/r_1'r_2'.$$

The product of $(p_1, q_1)/r_1$ and $(p_2, q_2)/r_2$ is defined to be

$$(p_1p_2 + q_1q_2, p_1q_2 + p_2q_1)/r_1r_2$$

and again we may readily verify that if

$$(p_1, q_1)/r_1 = (p_1', q_1')/r_1' \text{ and } (p_2, q_2)/r_2 = (p_2', q_2')/r_2'$$

then

$$(p_1, q_1)/r_1 \cdot (p_2, q_2)/r_2 = (p_1', q_1')/r_1' \cdot (p_2', q_2')/r_2'.$$

Defining subtraction in terms of addition we have

$$(p_1, q_1)/r_1 - (p_2, q_2)/r_2 = (p_3, q_3)/r_3 \leftrightarrow (p_1, q_1)/r_1 = (p_2, q_2)/r_2 + (p_3, q_3)/r_3$$

so that

$$(p_1, q_1)/r_1 - (p_2, q_2)/r_2 = (p_1, q_1)/r_1 + (q_2, p_2)/r_2,$$

and for division we have, if $p_2 \neq q_2$,

$$(p_1, q_1)/r_1 \div (p_2, q_2)/r_2 = (p_3, q_3)/r_3 \leftrightarrow (p_1, q_1)/r_1 = (p_2, q_2)/r_2 \cdot (p_3, q_3)/r_3$$

so that

$$(p_1, q_1)/r_1 \div (p_2, q_2)/r_2 = (p_1, q_1)/r_1 \cdot (p_2r_2, q_2r_2)/|p_2, q_2|^2$$

where $|p, q|$ is the positive difference of p and q.

The familiar abbreviations $+s/r$, $-s/r$ are introduced by the equations

$$+s/r = (p+s, p)/r, \quad -s/r = (p, p+s)/r$$

and as usual $+p/1$, $-p/1$ are denoted by $+p$, $-p$ respectively and $(p, p)/r$, which is equal both to $+0/r$ and to $-0/r$, is denoted simply by 0.

We shall use the letters k, m, n, p, q, r, with or without suffixes to denote natural numbers; x and y stand for rational numbers. We shall be chiefly concerned with functions of a rational variable x, denoted by $f(x)$, $g(x)$ etc., and with functions of a rational variable x and a natural number variable n, denoted by $f(n, x)$, $g(n, x)$ etc.

1, 2. 1 It might be thought that the class of rational recursive functions could be increased by applying the original recursion schema to rational recursive functions of a natural number argument, but this is not the case.

For instance, if we define the rational $f(n, x)$ by the recursion

$$f(0, x)=0, \ f(Sn, x)=\varphi(n, x, f(n, x)) \tag{i}$$

where

$$\varphi(n, (p, q)/r, \ (u, v)/w)=(a, b)/c,$$

a, b and c being recursive functions of n, p, q, r, u, v, w, then we can find recursive functions $u_n(p, q, r)$, $v_n(p, q, r)$, $w_n(p, q, r)$ such that

$$f(n, (p, q)/r)=(u_n, v_n)/w_n \tag{ii}$$

It suffices to define u_n, v_n, w_n by the simultaneous recursions $u_0=v_0=0$, $w_0=1$ and

$$u_{n+1}(p, q, r)=a(n, u_n, v_n, w_n)$$
$$v_{n+1}(p, q, r)=b(n, u_n, v_n, w_n)$$
$$w_{n+1}(p, q, r)=c(n, u_n, v_n, w_n);$$

then u_n, v_n, w_n are recursive, and therefore $f(n, x)$ defined by (ii) is recursive, and satisfies the recursion (i).

1, 3 Notation

For natural numbers p, q with $q>0$, we denote by $[p/q]$ the quotient of p divided by q; for integers i, j with $j\neq 0$, we define

$$[i/j]=[|i|/|j|] \text{ if } i\,j\geqslant 0$$
$$=-[|i|/|j|] \text{ if } i\,j<0.$$

If for some integer k and rational x, $[10^k x]=0$ we write $x=0(k)$ so that $x=0(k)$ is equivalent to the recursive relation $|x|<1/10^k$.

1, 3. 1 A relation which holds for all sufficiently great values of n is said to hold for majorant n. More specifically a relation

$$R(n, m_1, m_2, \ldots, m_p)$$

holds *for majorant n* if there is a recursive function $M(m_1, m_2, \ldots, m_p)$ such that

$$n \geqslant M(m_1, m_2, \ldots, m_p) \rightarrow R(n, m_1, m_2, \ldots, m_p)$$

with free variables $m_1, m_2, \ldots, m_p$, where M reduces to a constant if there are no variables in R beyond n.

More generally, $R(m, n, r_1, r_2, \ldots, r_p)$ holds *for majorant m, n*, (the order of m, n being material) if there are recursive functions

$$M(r_1, r_2, \ldots, r_p),\ N(m, r_1, r_2, \ldots, r_p)$$

such that

$$m \geqslant M(r_1, r_2, \ldots, r_p) \,\&\, n \geqslant N(m, r_1, r_2, \ldots, r_p) \rightarrow R(m, n, r_1, r_2, \ldots, r_p).$$

1, 3. 2 *Relative equality*

If $f(n)-g(n)=0(k)$ for majorant n, we say that $f(n)$ is equivalent to $g(n)$ and write

$$f(n)=g(n), \text{ relative to } n.$$

Similarly if $f(n)>-10^{-k}$ for majorant n, we write

$$f(n) \geqslant 0, \text{ relative to } n$$

and if $f(n)<10^{-k}$ for majorant n, then

$$f(n) \leqslant 0, \text{ relative to } n.$$

Furthermore if

$$f(m, n)-g(m, n)=0(k)$$

for majorant m, n then we write

$$f(m, n)=g(m, n), \text{ relative to } m, n.$$

1, 3. 3 *Recursive convergence*

A primitive (general) recursive function $f(n)$ is primitively (general) *recursively convergent* if there is a primitive (general) recursive function $N(k)$ such that $N(k+1) \geqslant N(k) \geqslant k$ and

$$N \geqslant n \geqslant N(k) \rightarrow f(N) - f(n) = 0(k).$$

When it is necessary to emphasize the connection of N with f we attach f to N as a suffix. We follow this practice systematically in the sequel, both in relation to the N of this definition and functions $c(k)$, $d(k)$ we meet later. Any parameters in f may also appear in N.

A recursive function $f(n, x)$ of a positive integral variable n and rational variable x, is recursively convergent in an interval $a \leqslant x \leqslant b$ if there is a recursive function $N(k, x)$ such that

$$a \leqslant x \leqslant b \,\&\, N \geqslant n \geqslant N(k, x) \rightarrow f(N, x) - f(n, x) = 0(k).$$

If there is a recursive function $N(k)$, independent of x, such that

$$a \leqslant x \leqslant b \,\&\, N \geqslant n \geqslant N(k) \rightarrow f(N, x) - f(n, x) = 0(k)$$

then $f(n, x)$ is said to be *uniformly recursively convergent* in (a, b). In particular if $N(k) = k$, $f(n, x)$ is uniformly recursively convergent in standard form.

Since

$$f(N(k), x) - f(n, x) = 0(k) \text{ for } n \geqslant N(k)$$

therefore $f(N(k), x)$ is equivalent to $f(n, x)$ and since

$$p > q \rightarrow f(N(p), x) - f(N(q), x) = 0(k)$$

it follows that $f(N(k), x)$ is a standard form equivalent of the uniformly recursively convergent function $f(n, x)$.

1, 3. 4 *Relative convergence*

A recursive function $f(m, n)$ is said to be convergent in m, relative to n, if there is a recursive $M(k)$ with $M(k+1) > M(k)$ such that,

$$m_1 \geqslant M(k) \,\&\, m_2 \geqslant M(k) \rightarrow f(m_1, n) - f(m_2, n) = 0(k),$$

for majorant n.

A function $f(m, n)$, convergent in m, relative to n, is not necessarily convergent for any fixed value of n. For instance m/n is convergent in m, relative to n, since

$$(m_1 - m_2)/n = 0(k)$$

if $m_1 = m_2$, and if $m_1 \neq m_2$ and $n > 10^k|m_1 - m_2|$; but for fixed n, and any m,

$$(M - m)/n > 1 \text{ if } M > m + n,$$

so that m/n is not convergent for fixed n.

1, 3. 5 *The reduced sequence*

Given any recursive function $f(m, n)$, recursively convergent in n and convergent in m relative to n, we can find a recursively convergent $R_f(m)$ such that

$$f(m, n) = R_f(m), \text{ relative to } m, n.$$

Since $f(m, n)$ is convergent in n, there is a recursive function $N(k, m)$ such that

$$n^* \geqslant n \geqslant N(k, m) \rightarrow f(m, n^*) - f(m, n) = 0(k),$$

and since $f(m, n)$ converges relative to n, there is a recursive $M(k)$ such that

$$m^* \geqslant m \geqslant M(k) \rightarrow f(m^*, n) - f(m, n) = 0(k),$$

for majorant n.

Let $R_f(p) = f(M(p), N(p, M(p)))$; $R_f(p)$ is called the *reduce* of $f(m,n)$. We show first that the reduce is recursively convergent.

For majorant n, we have

$$f(M(p), N(p, M(p))) - f(M(p), n) = 0(p),$$
$$f(M(q), N(q, M(q))) - f(M(q), n) = 0(q)$$

and, if $q > p$,

$$f(M(p), n) - f(M(q), n) = 0(p)$$

so that, for $q > p$,

$$R_f(p) - R_f(q) = 3 \cdot 0(p)$$

which proves convergence. It remains to show that

$$f(m, n) = R_f\ (m), \text{ relative to } m, n.$$

We observe that

$$f(m, n) - f(M(m), n) = 0(k), \text{ for majorant } m, n,$$

and

$$f(M(m), n) - f(M(m), N(m, M(m))) = 0(m), \text{ for majorant } n,$$

whence $f(m, n) - R_f(m) = 2 \cdot 0(k)$, for majorant m, n as was to be shown.

1, 4 The familiar theorem of classical analysis that a monotone bounded sequence is convergent does not hold in recursive analysis. To prove this we shall utilise an important theorem in the metamathematics of arithmetic; this theorem affirms that there can be no effective procedure for deciding of any primitive recursive function $f(n)$ whether there is value for n of which $f(n) > 0$, or $f(n) = 0$ for all n. A proof of this theorem is given in the Hilbert-Bernays Grundlagen (Ref. 3 Vol. 2, p. 417–8).

Let $f(n)$ be any primitive recursive function and let

$$F(0) = 0,\ F(n+1) = F(n) + [1 \dot{-} \{F(n) + (1 \dot{-} f(n))\}]$$

so that $F(n)$ takes only the values 0,1 and is non-decreasing. If it were true that a monotone bounded sequence were recursively convergent then there would be a primitive recursive $N(k)$ such that

$$n \geqslant N(k) \rightarrow F(n) - F(N(k)) = 0(k),$$

and in particular

$$n \geqslant N(1) \rightarrow |F(n) - F(N(1))| < 1,$$

so that, since $F(n)$ is an integer, $F(n) = F(N(1))$ for $n \geqslant N(1)$. If $F(N(1)) = 0$ it follows that $F(n) = 0$ for $n \geqslant N(1)$ and since $F(n)$ is non-decreasing, therefore $F(n) = 0$ also for $n < N(1)$. Thus if $F(N(1)) = 0$, then $F(n) = 0$ for all n; however if $f(n) = 1$ for some n then $F(n+1) = 1$, and if $f(n) = 0$ for all n then $F(n) = 0$ for all n. It follows that if $F(N(1)) = 0$ then $f(n) = 0$ for all n and if $F(N(1)) = 1$ then $f(n) = 1$ for some $n \leqslant N(1)$. Since $F(n)$ and $N(k)$ are recursive, the value of $F(N(1))$ is computable in a finite number of steps and provides therefore an effective procedure for deciding whether

$f(n)=0$ for all n, or not. Since no effective decision procedure is possible the hypothesis that $F(n)$ is recursively convergent is untenable. Many of the classical tests for convergence are, however, valid tests for recursive convergence, as we shall show.

1, 4. 1 *Limits*

A rational number l is the recursive limit of the recursively convergent sequence $s(n)$, and $s(n)$ is said to tend recursively to the limit l if there is a recursive function $n(k)$ such that

$$n \geqslant n(k) \rightarrow l - s(n) = 0(k).$$

It follows, of course, that if $s(n)$ tends recursively to a limit, then $s(n)$ is recursively convergent.
A recursive function $s(n)$ is *recursively divergent* if there is a recursive $d(n)$ and an integer k such that

$$|s(n+d(n)) - s(n)| \geqslant 10^k.$$

The condition for a sequence $s(n)$ *not* to tend recursively to a limit is that there should be recursive functions

$$\nu(p, q, r),\ n(p, q, r, k) \text{ such that } n(p, q, r, k) \geqslant k$$

and

$$n = n(p, q, r, k) \rightarrow |s(n) - (p, q)/(r+1)| \geqslant 10^{-\nu(p, q, r)}.$$

To illustrate this condition we prove that the sequence $s(n) = u(n)/n!$, where $u(0)=1$, $u(n+1)=(n+1)u(n)+1$, is recursively convergent, but does not tend recursively to a limit. For

$$n > q \rightarrow 0 < s(n) - s(q) < 1/q(q!)$$

so that $s(n)$ is recursively convergent; but (ignoring, as we may, negative rationals)

$$\begin{aligned}\{s(q)-p/q\}q! &= u(q) - p\{(q-1)!\}\\ &= q(q-1)u(q-2) - p\{(q-1)!\} + q + 1\\ &= 2(\mathrm{mod}\ q-1),\ q \geqslant 4.\end{aligned}$$

Thus

$$|s(n) - p/q| \geqslant (1-1/q)/q!,$$

for all n such that $n>q\geqslant 4$; in the cases $q=1, 2, 3$, with $n>4$, the difference $|s(n)-p/q|$ is least for $p=8, q=3$ and $|s(n)-8/3|>1/24$, which completes the proof that $s(n)$ does not tend recursively to p/q for any p, q.

1, 4. 2 *Tests for recursive convergence*

We commence by proving the recursive convergence of the geometric series Σx^n for $0\leqslant x<1$. From

$$nx^n < \sum_{r\leqslant n} x^r = (1-x^{n+1})/(1-x) < 1/(1-x)$$

it follows that $x^n < 1/n(1-x)$ and so x^n tends recursively to zero and $\sum_{r\leqslant n} x^r$ tends recursively to $1/(1-x)$.

If $x\geqslant 1$, $\sum_{r\leqslant n+1} x^r - \sum_{r\leqslant n} x^r = x^{r+1}\geqslant 1$, so that $\sum_{r\leqslant n} x^r$ is recursively divergent. And similarly if $a(n)$ does *not* tend recursively to zero, so that $a(n+1)\geqslant 10^{-\nu(0,0,0)}$ when $n+1=n(0, 0, 0, k)$, then

$$\sum_{r\leqslant n+1} a(r) - \sum_{r\leqslant n} a(r) = a(n+1) \geqslant 10^{-\nu(0,0,0)}$$

when $n+1 = n(0, 0, 0, k)$, showing that $\sum_{r\leqslant n} a(r)$ is recursively divergent. Furthermore if $\sum_{r\leqslant n} a(r)$ is recursively divergent, and $a(r)>0$, then, for any fixed m, $1/\sum_{m\leqslant r\leqslant n} a(r)$ tends recursively to zero; let $s(n)=\sum_{m\leqslant r\leqslant m+n} a(r)$ and $e(0)=0$, $e(n+1)=e(n)+d(e(n))$, so that $s(e(n+1))-s(e(n))\geqslant 10^k$, by hypothesis, from which it follows that $s(e(n))\geqslant n\cdot 10^k$ and therefore

$$p\geqslant e(n) \rightarrow 1/s(p)\leqslant 1/s(e(n))\leqslant 10^{-k}/n$$

showing that $s(n)$ is recursively convergent to zero.

1, 4. 3 The comparison tests for positive series are obviously valid for recursive convergence, for if $u(n)\leqslant kv(n)$ then

$$\sum_{m\leqslant r\leqslant m+n} u(r) \leqslant k \sum_{m\leqslant r\leqslant m+n} v(r)$$

and so if $\Sigma v(r)$ is recursively convergent then so is $\Sigma u(r)$, and if $\Sigma u(r)$ is recursively divergent so too is $\Sigma v(r)$.

Moreover, from the inequality

$$u(n+1)/u(n) \leqslant v(n+1)/v(n)$$

follows $u(n) \leqslant \{u(0)/v(0)\}v(n)$ so that the recursive convergence of $\Sigma v(n)$ entails that of $\Sigma u(n)$, and $\Sigma v(n)$ is recursively divergent with $\Sigma u(n)$.

1, 4. 4 For Cauchy's test, we observe that if for some fixed k

$$u(n) \leqslant k^n < 1$$

then $\Sigma u(n)$ is recursively convergent by the foregoing tests, and if $u(n) \geqslant 1$ for $n = n(r)$, where $n(r)$ is strictly increasing, then it follows in turn that $u(n)$ does not tend recursively to zero and $\Sigma u(n)$ diverges.

1, 4. 5 For D'Alembert's test, we note that from

$$u(n+1)/u(n) \leqslant k < 1$$

for a fixed k, follows $u(n+1)/u(n) \leqslant k^{n+1}/k^n$, proving that $\Sigma u(n)$ converges recursively, and from

$$u(n+1)/u(n) \geqslant 1$$

follows the recursive divergence of $\Sigma u(n)$, since $u(n) \geqslant u(0)$.

1, 4. 6 Kummer's test requires closer consideration. The original form of the test is valid for recursive convergence, for if there is a constant $\alpha > 0$ such that

$$c(n)\{u(n)/u(n+1)\} - c(n+1) \geqslant \alpha$$

and if $c(n)u(n)$ tends recursively to zero, then $\Sigma u(r)$ converges recursively since

$$\alpha \sum_{n+1 \leqslant r \leqslant N} u(r) \leqslant c(n)u(n) - c(N)u(N) < c(n)u(n).$$

Dini's form of the test, which omits the condition that $c(n)u(n)$ tends recursively to zero, is not valid, since it is equivalent to asserting the recursive convergence of a monotone bounded sequence; for if $s(n)$ is any strictly increasing recursive function bounded above by an integer B and if

$$u(n) = s(n) - s(n-1),\ v(n) = B - \sum_{1 \leqslant r \leqslant n} u(r)$$

and $c(n)=v(n)/u(n)$ then

$$c(n)\{u(n)/u(n+1)\}-c(n+1)=1.$$

1, 4. 7 We consider next Cauchy's condensation test. Denote $\sum_{m\leqslant r\leqslant n} f(r)$ by $f(m, n)$; then if $f(r)$ is steadily decreasing, $f(2^n, 2^{n+1}-1)<2^n f(2^n)$, and therefore

$$f(2^n, 2^{N+1}-1)<\sum_{n\leqslant r\leqslant N} 2^r f(2^r),$$

whence it follows that, if $2^{N+1}\geqslant M+1$,

$$\sum_{2^n\leqslant r\leqslant M} f(r)<\sum_{n\leqslant r\leqslant N} 2^r f(2^r).$$

Similarly, if $2^n+1\geqslant m$,

$$\sum_{m\leqslant r\leqslant 2^{N+1}} f(r)\geqslant\sum_{n\leqslant r\leqslant N} 2^r f(2^{r+1}).$$

These inequalities show that the condensation test is valid for recursive convergence and divergence. We deduce that $\Sigma\, 1/r$ is recursively divergent so that $1/\sum_{m\leqslant r\leqslant m+n}(1/r)$ tends recursively to zero, for any fixed m. Furthermore $\Sigma\, 1/r^2$ is recursively convergent with $\Sigma\, 1/2^r$.

1, 4. 8 Raabe's tests for recursive convergence and divergence take the form:

If for some $k>1$,

$$u(n)/u(n+1)\geqslant 1+k/n$$

then $\Sigma u(n)$ is recursively convergent, and if

$$u(n)/u(n+1)\leqslant 1+1/n+1/n\sum_{1\leqslant r\leqslant n}(1/r)$$

then $\Sigma\, u(n)$ is recursively divergent.

For the convergence test we observe that the given inequality implies that

$$\{n\cdot u(n)/u(n+1)\}-(n+1)\geqslant k-1>0;$$

from this we deduce that

$$n\cdot u(n)/(n+1)\cdot u(n+1)\geqslant 1+(k-1)/(n+1)$$

and hence, by multiplication,

$$\text{(using } a>0 \,\&\, b>0 \rightarrow (1+a)(1+b)>a+b)$$

we find

$$N\cdot u(N) \leqslant n\cdot u(n)/(k-1) \sum_{n+1\leqslant r\leqslant N} (1/r)$$

so that, keeping n fixed, $N\cdot u(N)$ tends recursively to zero, and $\Sigma u(n)$ converges recursively, by Kummer's test. For the divergence test we start by proving that if

$$v(r)=1/r \sum_{1\leqslant p\leqslant r} (1/p)$$

then $\Sigma v(r)$ is recursively divergent. For $1/2^n v(2^n)<n+1$, so that $\Sigma\, 2^n v(2^n)$ diverges, by comparison with $\Sigma\, 1/(n+1)$, and hence, by the condensation test, $\Sigma v(r)$ itself is recursively divergent. Since

$$v(n)/v(n+1)=\{1+(n+1) \sum_{1\leqslant r\leqslant n} (1/r)\}/n \sum_{1\leqslant r\leqslant n} (1/r)$$

$$=1+1/n+1/n \sum_{1\leqslant r\leqslant n} (1/r) \geqslant u(n)/u(n+1),$$

therefore $\Sigma u(n)$ is recursively divergent.

In preparation for a proof of the Gauss test for recursive convergence we shall prove

1, 4. 9 If $\Sigma u(r)$ is recursively convergent, and $n\cdot u(n)$ is steadily decreasing and tends recursively to zero, then $\{\sum_{1\leqslant r\leqslant n} 1/r\}\, n\cdot u(n)$ tends recursively to zero.

For $n\geqslant r$ we have

$$n\cdot u(n)/r=\{n\cdot u(n)/r\cdot u(r)\}u(r)\leqslant u(r)$$

and since $\Sigma u(r)$ is recursively convergent there is a recursive $n(k)$ such that, writing $h(m, n)$ for $\sum_{m\leqslant r\leqslant n} (1/r)$,

$$n>n(k) \rightarrow h(n(k)+1, n)\cdot nu(n)<1/2k.$$

Further, since $n\cdot u(n)$ tends recursively to zero there is a recursive $r(k)$ such that $h(1, n(k))\cdot nu(n)<1/2k$, for $n>r(k)$, whence it follows that

$$n>max(n(k), r(k)) \rightarrow h(1, n)\cdot nu(n)<1/k.$$

The foregoing tests may be summarised in

The Gauss test. If there are constants α, β, M and a recursive $\theta(n)$ such that $|\theta(n)| < M$ and

$$u(n)/u(n+1) = \alpha + \beta/n + \theta(n)/n^2$$

then

$\alpha > 1$, or $\alpha = 1$ and $\beta > 1$, implies $\Sigma u(n)$ is recursively convergent and

$\alpha < 1$, or $\alpha = 1$ and $\beta \leqslant 1$, implies $\Sigma u(n)$ is recursively divergent.

The case $\alpha = 1, \beta \leqslant 1$ alone requires further consideration. By the previous theorem (taking $1/n^2$ for $u(n)$) we know that $(1/n) \sum_{1 \leqslant r \leqslant n} (1/r)$ tends recursively to zero, and so there is a recursive $N(k)$ such that

$$n \geqslant N(M) \rightarrow \frac{\theta(n)}{n^2} < \frac{1}{n \sum_{1 \leqslant r \leqslant n} (1/r)}$$

which shows that $\alpha = 1, \beta \leqslant 1$ suffices to ensure the recursive divergence of $\Sigma u(r)$.

1, 5 We turn to consider the tests for relative convergence.

Relative convergence has already been defined. A recursive function $s(m, n)$ is said to *divergent in* n, *relative to* m, if there is a constant $k \geqslant 1$, and a recursive function $\lambda(n)$ such that

$$|s(m, n+\lambda(n)) - s(m, n)| \geqslant 1/k,$$

for majorant n, m.

$s(m, n)$ is said to *tend to zero, relative to* m, if

$$s(m, n) = 0(k)$$

for majorant n, m.

In the following theorems $a(m, n)$, $b(m, n)$, $c(m, n)$ are positive, non-zero, recursive functions.

THEOREM 1 *If for some fixed* k, *and majorant* m,

$$a(m, n) \leqslant k < 1$$

then $\sum_{0 \leqslant r \leqslant n} \{a(m, r)\}^r$ *is convergent in* n, *relative to* m.

For $\sum_{n \leqslant r \leqslant N} \{a(m, r)\}^r \leqslant \sum_{n \leqslant r \leqslant N} k^r$, for majorant m. If $i(r+1) \geqslant r+i(r)$ and if $a(m, i(r)) \geqslant 1$ for majorant m, then $\sum_{0 \leqslant r \leqslant n} \{a(m, r)\}^r$ diverges relative to m. For

$$\sum_{0 \leqslant r \leqslant i(n+1)} \{a(m, r)\}^r - \sum_{0 \leqslant r \leqslant n} \{a(m, r)\}^r \geqslant 1,$$

for majorant m.

Theorem 1, 1 *If for a fixed* $\lambda > 0$,

$$a(m, n) \leqslant \lambda\, b(m, n)$$

for majorant m, *then the convergence of* $\Sigma\, b(m, r)$ *relative to* m, *entails that of* $\Sigma\, a(m, r)$ *relative to* m, *and the relative divergence of he latter entails that of the former.*

For

$$\sum_{n \leqslant r \leqslant N} a(m, r) \leqslant \lambda \sum_{n \leqslant r \leqslant N} b(m, r)$$

for majorant m.

Theorem 1, 1. 1 *If there are constants* h, H *such that*

$$0 < h \leqslant a(m, 0) \leqslant H,\ h \leqslant b(m, 0) \leqslant H$$

and $a(m, n+1)/a(m, n) \leqslant b(m, n+1)/b(m, n)$, *for majorant* m, *then the relative convergence of* $\sum_{r \leqslant n} b(m, r)$ *entails that of* $\sum_{r \leqslant n} a(m, r)$, *and the relative divergence of the latter entails that of the former.*

For $a(m, n) \leqslant (H/h)\, b(m, n)$, for majorant m.

Theorem 1, 1. 2 *The condition* $a(m, n+1)/a(m, n) \leqslant k < 1$, *for majorant* m, *is sufficient for relative convergence of* $\Sigma a(m, n)$, *since* $a(m, n+1)/a(m, n) \leqslant k^{n+1}/k^n$, *for majorant* m.
The condition $a(m, n+1)/a(m, n) \geqslant 1$, *for majorant* m, *suffices for relative divergence, since* $\sum_{r \leqslant n} 1$ *diverges.*

Theorem 1, 1. 3 *If there is a constant* $\beta > 0$ *such that*

$$c(m, n)\{a(m, n)/a(m, n+1)\} - c(m, n+1) \geqslant \beta,$$

for majorant m, and if $c(m,n)a(m,n)$ tends to zero in n, relative to m, then $\sum_{r\leqslant n} a(m,r)$ *converges relative to m.*

For $\beta \sum_{n+1\leqslant r\leqslant N} a(m,r) \leqslant c(m,n)a(m,n)$, m majorant.

THEOREM 1, 1. 4 *If there is a constant $p>1$ such that*

$$a(m,n)/a(m,n+1) \geqslant 1+p/n,$$

for majorant m, then $\sum_{r\leqslant n} a(m,r)$ *converges relative to m.*

We readily prove $n\cdot a(m,n)$ tends to zero in n, relative to m, and the result then follows from the previous theorem.

THEOREM 1, 1. 5 *If* $a(m,n)/a(m,n+1) \leqslant 1+1/n+1/n \sum_{r=1}^{n} (1/r)$, *for majorant m, then* $\sum_{r\leqslant n} a(m,r)$ *diverges relative to m.*

(The proof is the same as for recursive divergence.)

THEOREM 1, 1. 6 *If $f(m,n)$ is monotonic decreasing in n, relative to m, i.e. if $f(m,N)<f(m,n)$, for $N>n$, and majorant m, then*

$$\sum_{r\leqslant n} f(m,r), \quad \sum_{r\leqslant n} 2^r f(m,2^r)$$

converge and diverge together, relative to m.

(The proof is the same as for recursive convergence.)

THEOREM 1, 1. 7 *If* $a(m,n)/a(m,n+1)=\alpha+\beta/n+\theta(m,n)/n^2$, *for majorant m, and if* $|\theta(m,n)|<M$, *then* $\alpha>1$, *or* $\alpha=1, \beta>1$, *are sufficient conditions for relative convergence, and* $\alpha<1$, *or* $\alpha=1, \beta\leqslant 1$ *are sufficient for relative divergence of* $\sum_{r\leqslant n} a(m,n)$.

This follows from Theorems 1, 1. 2; 1, 1. 4; 1, 1. 5.

We proceed next to sharpen Theorem 1, 1. 7, proving first a series of preparatory theorems.

THEOREM 1, 1. 8 *If $a>b$, and $n\geqslant 2$, then*

$$nb^{n-1} < (a^n-b^n)/(a-b) < na^{n-1}.$$

PROOF. If $x>1, n>1$, then $\sum_{r\leqslant n} x^r > n+1$, and therefore

$$(x^{n+1}-1)/(x-1) > n+1,$$

whence, taking $x=a/b$,

$$(a^{n+1}-b^{n+1})/(a-b)>(n+1)b^n.$$

Similarly, if $x<1, n>1$, then $\sum_{r\leqslant n} x^r<(n+1)$ and therefore

$$(1-x^{n+1})/(1-x)<n+1,$$

and the second half of the theorem follows by taking $x=b/a$.

We note that since $t^q-1=\{\sum_{r\leqslant q-1} t^r\}(t-1)$, it follows that if $t>0$, then $t \gtreqless 1$ according as $t^q \gtreqless 1$.

THEOREM 1, 1. 8. 1 *If α is positive, and p, q are positive integers, $p>q$, then $(1+\alpha)^p>\{1+(p/q)\alpha\}^q$.*

This is a simple consequence of a special case of the famous theorem of the means, viz. $\{(ma+nb)/(m+n)\}^{m+n}>a^m b^n$, $a\neq b$, m, n positive, non-zero integers. (There are several algebraic proofs of the theorem of the means which are valid in the present system.)

Take $m=q$, $m+n=p$, $a=1+(p/q)\alpha$, $b=1$ and the theorem follows. Similarly, if $p<q$, taking $m=p$, $m+n=q$, $b=1$, $a=1+\alpha$, we have $(1+\alpha)^p<\{1+(p/q)\alpha\}^q$.

THEOREM 1, 1. 9 *If p, q, x are integers, $x>1, q>1, p\geqslant 1$, and if x_k is the greatest integer such that $(x_k)^q\leqslant x^p\cdot 2^{kq}$, then $x_n/2^n$ is recursively convergent.*

PROOF. Since $(2x_k)^q\leqslant x^p\cdot 2^{(k+1)q}$, therefore $x_{k+1}\geqslant 2x_k$ and so $x_k/2^k$ is monotonic increasing, and since x_0 is the greatest integer such that $(x_0)^q\leqslant x^p$, and $1^q<x^p$, then $x_0\geqslant 1$ and so $x_k>2^k$. From

$$(x_k)^q\leqslant x^p\cdot 2^{kq}<x^{pq}\cdot 2^{kq}$$

it follows that $x_k<x^p\cdot 2^k$. Hence

$$0\leqslant x^p-(x_k/2^k)^q<\{(x_k+1)^q-(x_k)^q\}/2^{kq}<q(x_k+1)^{q-1}/2^{kq}$$

(by Theorem 1, 1. 8)

$$<q\{x^p+1/2^k\}^{q-1}/2^k\leqslant q\{x^p+1\}^{q-1}/2^k$$

and so $(x_k/2^k)^q$ converges recursively to x^p.

Writing $x_k/2^k = y_k$, so that y_k is monotonic increasing and $y_k{}^q$ converges recursively, then

$$\{(y_{k+r})^q - (y_k)^q\}/\{y_{k+r} - y_k\} > q(y_k)^{q-1} > q,$$

i.e. $y_{k+r} - y_k < \{(y_{k+r})^q - (y_k)^q\}/q$, so that y_k is recursively convergent, which completes the proof.

1, 5. 1 We define, for $p \geqslant 1, q > 1, x > 1, \chi(x, p, q, k) = x_k/2^k$ (where x_k is given above) and $\chi(1, p, q, k) = \chi\ (0, p, q, k) = 1$.

THEOREM 1, 2 *$\sum_{r \leqslant n} 1/\chi(r, p, q, k)$ is convergent in n, relative to k, if $p > q$ and divergent in n, relative to k, if $p \leqslant q$.*

Since $\{\chi(x, p, q, k)\}^q$ converges recursively to x^p, therefore, writing $a_k = \chi(x+1, p, q, k)/\chi(x, p, q, k)$, we have if $p > q$, $(a_k)^q$ converges recursively to $(1+1/x)^p > (1+p/qx)^q$ by Theorem 1, 1. 8. 1; the recursive convergence of $a_k{}^q$ determines a recursive function $Q(x)$ such that $|(1+1/x)^p - (a_k)^q| < (1+1/x)^p - (1+p/qx)^q$ for $k \geqslant Q(x)$, and therefore $(a_k)^q > (1+p/qx)^q$, so that $a_k > 1 + p/qx$ for $k \geqslant Q(x)$. It follows, by Theorem 1, 1. 4, that $\sum_{r \leqslant n} 1/\chi(r, p, q, k)$ converges relative to k, if $p > q$.

If $p = q$, $\chi(n, p, q, k) = n$, and so $\sum_{r \leqslant n} 1/\chi(r, p, q, k)$ diverges. If $p < q$, $a_k < 1 + p/qx$ for $k \geqslant Q^*(x)$, whence, by Theorem 1, 1. 5, $\sum_{r \leqslant n} 1/\chi(r, p, q, k)$ diverges relative to k.

THEOREM 1, 2. 1 *If $\varphi(n, p, q, k) = n/\chi(n, p, q, k)$ then, for $p > q$, $\varphi(n, p, q, k)$ is monotonic decreasing to zero, relative to k.*

For

$$\varphi(n, p, q, k)/\varphi(n+1, p, q, k) > \{n/(n+1)\}\{1 + p/qn\}$$
$$= 1 + ((p/q) - 1)/(n+1)$$

if $k \geqslant Q(n), p > q$, so that $\varphi(n, p, q, k)$ is monotonic decreasing, relative to k, and further, if $N > n$,

$$\varphi(N, p, q, k)/\varphi(n, p, q, k) < 1/\prod_{N \geqslant r > n} \{1 + ((p/q) - 1)/r\}$$
$$< 1/((p/q) - 1) \sum_{N \geqslant r > n} (1/r) \text{ for majorant } k,$$

which shows that $\varphi(N, p, q, k)$ tends to 0 relative to k.

THEOREM 1, 2. 2 *If* $p>q, \{\sum_{1\leqslant r\leqslant n}(1/r)\}\, n/\chi(n, p, q, k)$ *tends to* 0, *relative to* k.

If $n>r$, by Theorem 1, 2. 1

$$\varphi(n, p, q, k)/\varphi(r, p, q, k)\chi(r, p, q, k)<1/\chi(r, p, q, k)$$

for majorant k. Since $\sum_{r\leqslant n} 1/\chi(r, p, q, k)$ converges relative to k, if $p>q$, given $x\geqslant 1$, we can determine $n(x)$ such that

$$\sum_{n(x)\leqslant r\leqslant n} 1/\chi(r, p, q, k)<1/2x,$$

for $n\geqslant n(x)$, and majorant k. Furthermore, since $\varphi(n, p, q, k)$ tends to 0, relative to k,

$$\{\sum_{1\leqslant r\leqslant n(x)}(1/r)\}\varphi(n, p, q, k)<1/2x,$$

for majorant n, k. Hence

$$\{\sum_{1\leqslant r\leqslant n}(1/r)\}\,\varphi(n, p, q, k)<1/x,$$

for majorant n, k, proving that $\{\sum_{1\leqslant r\leqslant n}(1/r)\}\,\varphi(n, p, q, k)$ tends to 0 relative to k.

1, 5. 2 We are now in a position to state the analogue, for relative convergence, of the Gauss test.

THEOREM 1, 3 *If for majorant* k, *and* $p>q$,

$$a(n, k)/a(n+1, k)=\alpha+\beta/n+\theta(n, k)/\chi(n, p, q, k),$$

and $|\theta(n, k)|<M$, *then* $\alpha>1$ *or* $\alpha=1, \beta>1$ *are sufficient conditions for the convergence of* $\sum_{r\leqslant n} a(r, k)$ *relative to* k, *and* $\alpha<1$ *or* $\alpha=1, \beta\leqslant 1$ *are sufficient conditions for relative divergence.*

We have to consider only the cases $\alpha=1, \beta>1$; $\alpha=1, \beta\leqslant 1$. If $\beta>1$, since $\theta(n, k)\varphi(n, p, q, k)$ tends to 0 relative to k, we can find N_0, such that

$$|n\theta(n, k)/\chi(n, p, q, k)|<(1/2)(\beta-1),$$

for $n\geqslant N_0$, and majorant k, and so

$$a(n, k)/a(n+1, k)>1+(1/2)(\beta+1)/n,$$

for $n \geqslant N_0$, and majorant k, which suffices for relative convergence, by Theorem 1, 1. 4.

Similarly, if $\beta < 1$, $a(n, k)/a(n+1, k) < 1 + (1/2)(\beta+1)/n$, which proves relative divergence, by Theorem 1, 1. 5.

If $\beta = 1$, since $\{\sum_{1 \leqslant r \leqslant n} (1/r)\}\, \varphi(n, p, q, k)$ tends to 0, relative to k, therefore we can find n_0 such that

$$|\theta(n, k)/\chi(n, p, q, k)| < 1/n \sum_{1 \leqslant r \leqslant n} (1/r),$$

for $n \geqslant n_0$, and majorant k, whence $\sum_{r \leqslant n} a(r, k)$ is divergent, relative to k, by Theorem 1, 1. 5.

1, 6 Recursive real numbers

A primitive recursive function $f(n)$ which is primitive recursively convergent is called a *primitive recursive real number*. A general recursive function $f(n)$ which is general recursively convergent is called a *general recursive real number*. When the distinction between primitive and general recursion is irrelevant (in the sense that the context allows either) we shall speak of a *recursive real number*.

If $f(n)$ and $g(n)$ are recursive real numbers and $f(n) = g(n)$, relative to n, we say that the recursive real numbers $f(n)$, $g(n)$ are equal, and write

$$\{f(n)\} = \{g(n)\}.$$

If for some constant $c > 0$, $f(n) > c$ for majorant n, we say that the recursive real number $f(n)$ is positive and write

$$\{f(n)\} > 0.$$

Similarly, if for a constant $c > 0$, $f(n) < -c$ for majorant n, then the recursive real number $f(n)$ is said to be negative and we write

$$\{f(n)\} < 0.$$

Two recursive real numbers $f(n)$, $g(n)$ are said to be recursively unequal if there is a constant $k > 0$, and a recursive $\lambda(n) \geqslant n$, such that

$$|f(\lambda(n)) - g(\lambda(n))| > k.$$

1, 6. 1 *Arithmetic of recursive real numbers*

If $s(n)$, $t(n)$ are recursively convergent then, patently, $s(n)+t(n)$ and $s(n)-t(n)$ are recursively convergent. The recursive real number $s(n)+t(n)$ is called the sum of the recursive real numbers $s(n)$ and $t(n)$ and we write

$$\{s(n)\}+\{t(n)\}=\{s(n)+t(n)\}.$$

The recursive real number $s(n)-t(n)$ is called the difference of $\{s(n)\}$ and $\{t(n)\}$ and we write

$$\{s(n)\}-\{t(n)\}=\{s(n)-t(n)\}.$$

A recursively convergent sequence is bounded, because there is a constant n_0 such that $|s(n)-s(n_0)|<1$, and so if

$$M=\max_{r\leqslant n_0}|s(n)|$$

and $S=M+1$, then $|s(n)|<S$.

If $s(n)$ is recursively convergent then $s(n)^2$ is recursively convergent for

$$|s(N)^2-s(n)^2|=|s(N)-s(n)|\cdot|s(N)+s(n)|<2S\cdot|s(N)-s(n)|.$$

It follows that if $s(n)$ and $t(n)$ are recursively convergent then so is

$$s(n)\cdot t(n)=\left\{\frac{s(n)+t(n)}{2}\right\}^2-\left\{\frac{s(n)-t(n)}{2}\right\}^2.$$

The recursive real number $s(n)\cdot t(n)$ is called the product of $\{s(n)\}$ and $\{t(n)\}$ and we write

$$\{s(n)\}\cdot\{t(n)\}=\{s(n)\cdot t(n)\}.$$

If $s(n)$ is recursively convergent, and is recursively different from 0 and if $s(n)\neq 0$, then

$$1/s(n) \text{ is recursively convergent.}$$

For there is a constant $k>0$ and a recursive $\lambda(n)\geqslant n$, such that

$$|s(\lambda(n))|>k$$

and a constant ν such that

$$n\geqslant\nu\ \&\ n_1\geqslant\nu\rightarrow|s(n)-s(n_1)|<\tfrac{1}{2}k$$

whence, since $\lambda(n_1) \geqslant n_1$ it follows that

$$n \geqslant n_1 \rightarrow |s(n)| > \tfrac{1}{2}k.$$

Since $s(n) \neq 0$, $\min\limits_{n \leqslant n_1} |s(n)| = m > 0$, and so, if $s = \min(m, \tfrac{1}{2}k)$, then

$$|s(n)| \geqslant s.$$

It follows that $1/s(n)$ is recursively convergent for

$$\left| \frac{1}{s(n)} - \frac{1}{s(N)} \right| \leqslant \frac{|s(N) - s(n)|}{s^2};$$

the recursive real number $1/s(n)$ is called the reciprocal of $\{s(n)\}$ and we write

$$1/\{s(n)\} = \{1/s(n)\}.$$

For recursive real numbers $\{s(n)\}$, $\{t(n)\}$ we define the quotient $\{s(n)\}/\{t(n)\}$ to be the real number $\{s(n)\} \cdot 1/\{t(n)\}$ when $\{t(n)\}$ is recursively different from zero.

THEOREM 1, 4 *There is no effective procedure for deciding of two recursive real numbers whether they are equal or not.*

Let $f(n)$ be any primitive recursive function, let $s(n) = 0$ and $t(0) = f(0)$, $t(n+1) = t(n) + f(n+1)$. Then $t(n) = 0$ if and only if $f(r) = 0$ for all $r \leqslant n$. Both $\{s(n)\}$ and $\{t(n)\}$ are primitive recursive real numbers. If there is an effective procedure for deciding whether $\{s(n)\} = \{t(n)\}$, or not, there is an effective procedure for deciding for an arbitrary $f(n)$, whether $f(n) = 0$ for all n, or there is an n for which $f(n) > 0$, and this we know to be impossible.

We define

$$\{f(n)\} > \{g(n)\} \leftrightarrow \{f(n)\} - \{g(n)\} > 0.$$

THEOREM 1, 5 *If $\{f(n)\}$ and $\{g(n)\}$ are recursively unequal then either $\{f(n)\} > \{g(n)\}$ or $\{g(n)\} > \{f(n)\}$.*

For there is a constant k and a strictly increasing recursive $\lambda(n)$ such that

$$|f(\lambda(n)) - g(\lambda(n))| > k;$$

by recursive convergence we can find N such that

$$m \geqslant n \geqslant N \rightarrow |f(m) - f(n)| < \tfrac{1}{3}k \;\&\; |g(m) - g(n)| < \tfrac{1}{3}k.$$

Hence if $f(\lambda(N)) > g(\lambda(N))$ then

$$f(\lambda(N)) - g(\lambda(N)) > k$$

and so

$$f(n) - g(n) > \tfrac{1}{3}k,$$

for $n \geqslant N$ proving that

$$\{f(n)\} > \{g(n)\}.$$

Similarly, if $f(\lambda(N)) < g(\lambda(N))$, then

$$\{g(n)\} > \{f(n)\}.$$

THEOREM 1, 5. 1 *If $\{f(n)\}$, $\{g(n)\}$ are unequal general recursive real numbers then they are general-recursively unequal.*

For if $\{f(n)\}$, $\{g(n)\}$ are unequal there is a constant k and, given any n, an integer N such that,

$$N \geqslant n \,\&\, |f(N) - g(N)| > k; \qquad (1)$$

let $\lambda(n)$ be the least value of N for which these inequalities hold then $\lambda(n)$ is general recursive, $\lambda(n) \geqslant n$, and $|f(\lambda(n)) - g(\lambda(n))| > k$, so that $\{f(n)\}$ and $\{g(n)\}$ are general-recursively unequal. It is important to observe that we have established only *general-recursive* inequality; even if $\{f(n)\}$ and $\{g(n)\}$ are unequal primitive recursive real numbers, we cannot prove primitive recursive inequality, since the existence of an N for which the inequalities (1) hold is no evidence for the existence of a primitive recursive $\lambda(n)$ for which

$$\lambda(n) \geqslant n \,\&\, |f(\lambda(n)) - g(\lambda(n))| > k.$$

A recursive real number $\{f(n)\}$ is rational if there is a rational number l such that $f(n)$ tends recursively to l, that is to say

$$f(n) - l = 0(k)$$

for majorant n.

A recursive real number $\{f(n)\}$ is *recursively irrational* if there are recursive functions $\nu(p, q, r) \geqslant 1$, $n(p, q, r)$ such that

$$n \geqslant n(p, q, r) \rightarrow |f(n) - (p, q)/(r+1)| \geqslant 1/\nu(p, q, r).$$

If the functions concerned are all primitive-recursive then $\{f(n)\}$ is primitive recursively irrational.

THEOREM 1, 6 *If a recursive real number $f(n)$ is irrational, that is to say if for any p, q, r, there exist k and m such that*

$$n \geqslant m \rightarrow |f(n)-(p, q)/(r+1)| \geqslant 10^{-k}$$

for all n, then $f(n)$ is general-recursively irrational.

For there is a recursive $n(s)$ such that,

$$n \geqslant n(s) \rightarrow |f(n)-f(n(s))| < 10^{-(s+1)},$$

and so by the irrationality of $\{f(n)\}$, there is a k such that

$$|f(n(k))-(p, q)/(r+1)| \geqslant 9 \cdot 10^{-(k+1)},$$

let $\varphi(p, q, r)$ be the least s such that

$$|f(n(s))-(p, q)/(r+1)| \geqslant 9 \cdot 10^{-(s+1)}$$

and let $N(p, q, r)=n(\varphi(p, q, r))$, so that $\varphi(p, q, r)$ and $N(p, q, r)$ are general recursive and

$$n \geqslant N(p, q, r) \rightarrow |f(n)-(p, q)/(r+1)| \geqslant 10^{-(\varphi(p, q, r)+1)}$$

which proves that $\{f(n)\}$ is general recursively irrational.

THEOREM 1, 6. 1 *There is no effective procedure for deciding whether any general recursive real number is rational or irrational.*

Let $s(0)=1$, $s(n+1)=1+(n+1)s(n)$ so that $\{s(n)/n!\}$ is the primitive recursive real number e, which of course is irrational. Let $f(n)$ be any primitive recursive function and let $t(n)=s(n)$ if $f(r)=0$ for all $r<n$, and $t(n)=s(k)$ if k is the first value of $r<n$ for which $f(r)>0$; for $N>n$, $t(N)/N!-t(n)/n! \leqslant s(N)/N!-s(n)/n!$ so that $t(n)/n!$ is primitive-recursively convergent.

If there exists an effective procedure for deciding whether $\{t(n)/n!\}$ is rational or irrational then there is an effective procedure for deciding whether or not there is an n for which $f(n) \neq 0$.

1, 7 *Recursive polynomials*

If a_n is a (primitive) recursive function,

$$\sum_{r \leqslant n} a_r x^r$$

is called *a (primitive) recursive polynomial of degree n*, with coefficients a_r, $r \leqslant n$.

If $\sum_{r \leqslant n} a_r t^n = 0$ for a certain t, then t is called a *zero* of the polynomial $\sum_{r \leqslant n} a_r x^r$.

If $f(x) = \sum_{r \leqslant n} a_r x^r$ has no rational zero and a_r, $r \leqslant n$, are integers, then

$$|f(p/q)| \geqslant 1/q^n.$$

For $q^n f(p/q) = \sum_{r \leqslant n} a_r p^r q^{n-r}$, and since $f(p/q) \neq 0$ and $\sum_{r \leqslant n} a_r p^r q^{n-r}$ is an integer, therefore

$$|\sum_{r \leqslant n} a_r p^r q^{n-r}| \geqslant 1$$

whence it follows that

$$|f(p/q)| \geqslant 1/q^n.$$

If $\sum_{r \leqslant n} a_r x^r$ has a rational zero p/q (p, q having no common factor) then p is a factor of a_0 and q is a factor of a_n; for

$$0 = q^n \sum_{r \leqslant n} a_r (p/q)^r = a_n p^n + q \sum_{r \leqslant n-1} a_r p^r q^{n-r-1}$$

so that a_n is divisible by q, and

$$0 = q^n \sum_{r \leqslant n} a_r (p/q)^r = a_0 q^n + p \sum_{1 \leqslant r \leqslant n} a_r p^{r-1} q^{n-r}$$

so that a_0 is divisible by p.

THEOREM 1, 7 *If $f_n(x)$ is the polynomial $\sum_{r \leqslant n} a_r x^r$, $M_n = \max_{r \leqslant n} |a_r|$ and z is at least as great as $|x|$, $|y|$ and unity, then*

$$\left| \frac{f_n(x) - f_n(y)}{x - y} \right| \leqslant \tfrac{1}{2} n(n+1) M_n z^{n-1}. \tag{1}$$

For

$$\frac{f_0(x)-f_0(y)}{x-y}=0$$

and if (1) holds for $n=k$ then

$$\left|\frac{f_{k+1}(x)-f_{k+1}(y)}{x-y}\right| = \left|\frac{f_k(x)-f_k(y)}{x-y}+a_{k+1}\frac{x^{k+1}-y^{k+1}}{x-y}\right|$$
$$\leqslant \tfrac{1}{2}k(k+1)\,M_k\,z^{k-1}+(k+1)\,M_{k+1}\,z^k$$
$$\leqslant \{\tfrac{1}{2}k(k+1)+(k+1)\}\,M_{k+1}\,z^k=\tfrac{1}{2}(k+1)(k+2)\,M_{k+1}\,z^k$$

so that (1) holds also for $n=k+1$, and therefore by induction (1) holds for all n.

THEOREM 1, 7. 1 *If $f(x)$ is a primitive recursive polynomial of degree n with integral coefficients a_r and no rational zeros, and if $\{s_n\}$ is a primitive recursive zero of $f(x)$, that is to say $\{s_n\}$ is a primitive recursive real number and $f(s_n)$ tends primitive recursively to zero then $\{s_n\}$ is primitive recursively irrational.*

Since s_n is recursively convergent we can find S such that $|s_n|\leqslant S$ for any n. Let

$$z=\max(|p/q|,S,1),$$
$$h=\max_{r\leqslant n}|a_r|, \text{ and } M=\tfrac{1}{2}n(n+1)hz^{n-1},$$

then, by the previous theorem, $|f(p/q)-f(s_m)|\leqslant M\cdot|p/q-s_m|$.

Since $f(s_m)$ tends recursively to zero we can find k so that

$$m\geqslant k\rightarrow|f(s_m)|<1/2q^n;$$

however,

$$|f(p/q)|\geqslant 1/q^n,$$

and therefore

$$|f(p/q)-f(s_m)|\geqslant 1/2q^n,$$

whence

$$m\geqslant k\rightarrow|p/q-s_m|\geqslant 1/2Mq^n$$

which proves that $\{s_m\}$ is primitive recursively irrational.

1, 7. 1 *Recursive expansions*

If $\{s_n\}$ is a (primitive) recursive real number and if a_n is a (primitive) recursive sequence such that

$$0 \leqslant a_{n+1} \leqslant r-1,\ r \geqslant 2,$$

and

$$\{s_n\} = \{\sum_{p \leqslant n} a_p r^{-p}\}$$

then the real number $\{s_n\}$ is said to have the (primitive) recursive expansion $\Sigma a_p r^{-p}$ in the scale of r.

THEOREM 1, 7. 2 *If $\{s_n\}$ is a recursively irrational recursive real number then $\{s_n\}$ has a recursive expansion in any scale $r \geqslant 2$.*

Since $\{s_n\}$ is a recursive real number there is a strictly increasing recursive function $n(k)$ such that

$$n \geqslant n(k) \rightarrow |s_n - s_{n(k)}| < 1/r^{k+1},$$

and since s_n is recursively irrational, there are recursive functions $i(p, q)$, $N(p, q)$ such that for any integers p, q, with $q > 0$,

$$n \geqslant N(p, q) \rightarrow |s_n - p/q| > r^{-i(p,q)}.$$

Taking $p=0, q=1$ we see that for all large enough j, k

$$|s_j - s_k| < r^{-i(0,1)}$$

and

$$|s_j| > r^{-i(0,1)},\ |s_k| > r^{-i(0,1)}$$

so that s_j, s_k are of the same sign. Without loss of generality we may therefore suppose that *all* s_n are positive.

Let $[x]$ denote the greatest integer contained in x, and let $p_k = [r^k s_{n(k)}]$; then *either*

$$0 \leqslant r^k s_{n(k)} - p_k \leqslant \tfrac{1}{2} \qquad \text{(i)}$$

or

$$r^k s_{n(k)} - p_k > \tfrac{1}{2}$$

so that

$$0 < p_k + 1 - r^k s_{n(k)} < \tfrac{1}{2}. \qquad \text{(ii)}$$

We consider first case (i).

For $n \geqslant n(k)$,

$$-\frac{1}{r} < r^k s_n - r^k s_{n(k)} < \frac{1}{r}$$

and so, by (i),

$$|r^k s_n - p_k| < 1;$$

but

$$n \geqslant N(p_k, r^k) \rightarrow |s_n - p_k/r^k| > r^{-i(p_k, r^k)}.$$

and therefore, for $n \geqslant N(p_k, r^k)$,

$$1 > |r^k s_n - p_k| > d_k = r^{k-i(p_k, r^k)}.$$

Taking $m(k) = \max\ (n(k), N(p_k, r^k))$, it follows that, for $n \geqslant m(k)$, each $r^k s_n$ lies either in the open interval $(p_k + d_k, p_k + 1)$ or in the open interval $(p_k - 1, p_k - d_k)$.

Writing $\mu(k) = n(i(p_k, r^k))$, we have

$$n \geqslant \mu(k) \rightarrow |r^k s_n - r^k s_{\mu(k)}| < d_k$$

which shows that all $r^k s_n$, for $n \geqslant \max\{\mu(k), m(k)\} = M(k)$, lie in one or other of these intervals and therefore

either $[r^k s_n] = p_k$, for all $n \geqslant M(k)$,

or $[r^k s_n] = p_k - 1$, for all $n \geqslant M(k)$,

and so $[r^k s_n] = [r^k s_{M(k)}]$, for $n \geqslant M(k)$.

In case (ii) the same result holds for $n \geqslant M^*(k)$, where

$$M^*(k) = \max\{n(i(p_k + 1, r^k)), n(k), N(p_k + 1, r^k)\}$$

and so if

$$\nu(0) = \max\{M(0), M^*(0)\}$$

and

$$\nu(k+1) = \max\{M(k+1), M^*(k+1), \nu(k) + 1\},$$

then

$$n \geqslant \nu(k) \rightarrow [r^k s_n] = [r^k s_{\nu(k)}],$$

and $\nu(k)$ is strictly increasing.

It follows that

$$0 \leqslant s_n - [r^k s_{\nu(k)}]/r^k < 1/r^k$$

for $n \geqslant \nu(k)$.

Let

$$a_0=[s_{\nu(0)}],\ a_{n+1}=[r^{n+1}s_{\nu(n+1)}]-r[r^n s_{\nu(n)}]$$

then

$$\sum_{k\leqslant n} a_k r^{-k}=[r^n s_{\nu(n)}]/r^n.$$

Now for any rational x,

$$0\leqslant rx—r[x]<r$$

and

$$-1<[rx]-rx\leqslant 0$$

so that

$$-1<[rx]-r[x]<r;$$

taking $r^n s_{\nu(n+1)}$ for x, it follows, since $[r^n s_{\nu(n)}]=[r^n s_{\nu(n+1)}]$, that

$$0\leqslant a_{n+1}<r,.$$

It remains to prove that the real numbers $\{s_n\}$ and $\{[r^n s_{\nu(n)}]/r^n\}$ are equal, for then $\Sigma a_k r^{-k}$ is a primitive recursive expansion of $\{s_n\}$ in the scale of r.

For $n\geqslant n(k+1)$,

$$|s_n-s_{\nu(n)}|<1/r^{k+1}$$

and

$$0\leqslant s_{\nu(n)}-[r^n s_{\nu(n)}]/r^n<1/r^n\leqslant 1/r^{k+1},$$

and therefore

$$|s_n-[r^n s_{\nu(n)}]/r^n|<1/r^k$$

for $n\geqslant n(k+1)$, which completes the proof.

In preparation for the last theorem of this chapter, that there is a primitive recursive real number whose decimal expansion is *not* primitive recursive, we recall some results from the theory of recursive functions.

By a theorem of Rosza Péter, all primitive recursive functions of one argument can be enumerated in the form

$$\varphi_0(n),\ \varphi_1(n),\ \varphi_2(n),\ \ldots$$

where

$$\varphi_0(n)=0,\ \varphi_1(n)=n+1,\ \varphi_2(n)=n-[\sqrt{n}]^2,\ \ldots.$$

In this enumeration each function of $\varphi_m(n)$ recurs infinitely often, the function $\varphi_p(n)$ and $\varphi_m(n)$ being identical for $p=3^m\cdot 7\cdot 11^x$, where x is any non-negative integer, and the function

$$\varphi_m(\varphi_m(n))=\varphi_p(n),$$

for $p=2\cdot 3^m\cdot 5^m$, so that $\varphi_m(n)$ and $\varphi_p(n)$ are different functions for $m>0$. It follows that the function $d(n)=\varphi_n(n)$ is *not* primitive recursive for if $d(n)$ were primitive recursive then so would $d(n)+1$ be, and therefore there would be a p such that

$$\varphi_p(n)=d(n)+1$$

identically, whence $\varphi_p(p)=d(p)+1=\varphi_p(p)+1$, which is impossible. The function $d(n)$ is however general recursive (being in fact definable by a double recursion) and so by the Kleene Normal Form Theorem, there is a primitive recursive function $\varphi(n)$ and a primitive recursive predicate $R(m, n)$ such that

$$d(n)=\varphi\{\mu_x R(n, x)\}$$

where $\mu_x R(n, x)$ denotes the least value of x for which $R(n, x)$ holds.

(The argument may be adapted to construct a function $d(n)$ which is general recursive but is not definable by multiple recursions of any finite order.)

By means of this function $d(n)$ we can now prove an important theorem due to E. Specker.

Theorem 1, 8 *There is a primitive recursive predicate $P(x)$, and a primitive recursive function $\gamma(n)$ which takes only the values* 0,1 *such that the general recursive function*

$$\gamma\{\mu_x(x\geqslant n\,\&\,P(x))\}$$

is not primitive recursive.

Let $\alpha(n)$, $\beta(n)$ be the primitive recursive functions with the properties

$$\alpha(0)=0,\ \alpha(n+1)=1;\ \beta(0)=1,\ \beta(n+1)=0;$$

and let $i(m, n)$, $j(m, n)$ be the representing functions of the primitive recursive predicates

$$(\exists x)\{n = 3^m \cdot 7 \cdot 11^x\},\ (\exists x)\{x \leqslant n \,\&\, R(m, x)\}.$$

Thus $i(m, n)$ and $j(m, n)$ take only the values 0,1 and $i(m, n)$ vanishes only if the functions $\varphi_m(t)$, $\varphi_n(t)$ are identical. In terms of i and j we define functions $a(n)$, $b(n)$, $c(n)$ by simultaneous recursion as follows:

$$a(0) = b(0) = c(0) = 0;$$

if $a(n) = 0$ then

$$a(n+1) = \beta(i(b(n), n+1))$$
$$b(n+1) = b(n) + a(n+1)$$
$$c(n+1) = n+1;$$

if $a(n) = 1$ then

$$a(n+1) = \alpha(j(c(n), n+1))$$
$$b(n+1) = b(n)$$
$$c(n+1) = c(n).$$

The functions $a(n)$, $b(n)$, $c(n)$ are primitive recursive; $a(n)$ takes only the values 0,1 since α and β take only these values. We shall show now that $a(n)$ vanishes infinitely often.

For suppose that, for certain n and p,

$$a(n) = 0,\ a(n+1) = a(n+2) = \ \dots\ = a(n+p) = 1;$$

then

$$c(n+r+1) = n+1,\ 0 \leqslant r \leqslant p,$$

and

$$b(n+r+1) = b(n)+1,\ 0 \leqslant r \leqslant p,$$

and

$$i(b(n)+1,\ n+r+1) = 0,\ 0 \leqslant r \leqslant p,$$

so that the functions $\varphi_{n+r+1}(t)$, $0 \leqslant r \leqslant p$. are all identical. However the functions $\varphi_{n+1}(t)$, $\varphi_{n+p+1}(t)$ are *not* identical if

$$n+p+1 = 2 \cdot 3^{n+1} \cdot 5^{n+1}$$

and so, for a large enough p, $a(n+p+1) = 0$.

On the other hand $a(n)=1$ for infinitely many n, for if

$$a(n)=a(n+1)=a(n+2)= \ldots =a(n+p)=0$$

then

$$b(n+r+1)=b(n),\ 0\leqslant r\leqslant p,$$

and

$$i(b(n),\ n+r+1)=1,\ 0\leqslant r\leqslant p,$$

so that none of the functions $\varphi_{n+r+1}(t)$, $0\leqslant r\leqslant p$, is identical with $\varphi_{b(n)}(t)$. Since however, if $m=3^{b(n)}\cdot 7\cdot 11^k$, then $\varphi_m(t)$ is identical with $\varphi_{b(n)}(t)$, and since $3^{b(n)}\cdot 7\cdot 11^k\geqslant 7\cdot 11^k>n$ for a large enough k, therefore $a(n+p+1)=1$ for a large enough value of p.

It follows that, for infinitely many values of n,

$$a(n)=0 \text{ and } a(n+1)=1$$

and so

$$b(n+1)=b(n)+1;$$

since $b(n)$ is non-decreasing, and changes by a unit at a time, this shows that the values of $b(n)$ comprise all the non-negative integers.

Let

$$\psi(n)=\varphi\{\mu_x(x\leqslant n\ \&\ R(c(n),\ x))\}$$

and

$$\gamma(n)=\beta(\psi(n)).$$

We shall prove that the function

$$g(n)=\gamma\{\mu_x(a(x)=0\ \&x\geqslant n)\}$$

is not primitive recursive.

For if $g(n)$ were primitive recursive we could find m so that

$$g(n)=\varphi_m(n)$$

identically; then since $m+1$ is a value of $b(t)$ for a suitable t, we can find a least t, $n+1$ say, such that $b(n+1)=m+1$ and therefore $b(n)=m$.

For this value of n we have

$$a(n)=0,\ \ a(n+1)=1,\ \ c(n+1)=n+1$$

and $i(m, n+1)=0$; this last equation shows that the functions $\varphi_m(t)$ and $\varphi_{n+1}(t)$ are identical and therefore, by hypothesis

$$g(n+1)=\varphi_{n+1}(n+1)=d(n+1).$$

We have seen that there is necessarily an $N>n$ for which $a(N)=0$; let $n+p+1$ be the least value (above n) for which $a(n+p+1)=0$, so that $a(n+r)=1$ for $1 \leqslant r \leqslant p$. Accordingly

$$\mu_x(a(x)=0 \,\&\, x \geqslant n+1)=n+p+1,$$
$$c(n+r+1)=n+1, \text{ for } 0 \leqslant r \leqslant p,$$

and

$$\begin{aligned}\psi(n+p+1)&=\varphi\{\mu_x(x \leqslant n+p+1 \,\&\, R(c(n+p+1), x))\}\\ &=\varphi\{\mu_x(x \leqslant n+p+1 \,\&\, R(n+1, x))\}.\end{aligned}$$

However, since $a(n+p)=1$ and $a(n+p+1)=0$, therefore

$$j(c(n+p), n+p+1)=1,$$

whence

$$j(n+1, n+p+1)=1,$$

which shows that there is a value of x such that $x \leqslant n+p+1$ and $R(n+1, x)$, and therefore

$$\psi(n+p+1)=\varphi\{\mu_x R(n+1, x)\}=d(n+1).$$

It follows that

$$\begin{aligned}g(n+1)&=\beta\{\psi(\mu_x(a(x)=0 \,\&\, x \geqslant n+1))\}\\ &=\beta\{\psi(n+p+1)\}=\beta\{d(n+1)\}\end{aligned}$$

and this contradiction establishes the Theorem.

THEOREM 1, 9 *There is a primitive recursive real number with a decimal expansion which is not primitive recursive.*

Let $a(n), \gamma(n)$ be the functions defined in the previous section, and let

$$\Gamma(n)=4\gamma(n)+1$$
$$\varphi(n)=3 \cdot \alpha(a(n))+\beta(a(n)) \cdot \Gamma(n).$$

Since $\gamma(n)$ takes only the values 0, 1 therefore $\Gamma(n)$ takes only the values 1, 5 and $\varphi(n)$ takes only the values 1, 3, 5. Furthermore since $a(n)=0$ for infinitely many n, therefore there are infinitely many values of n for which $\varphi(n)\neq 3$. We observe that the condition $\varphi(n)\neq 3$ is equivalent to $a(n)=0$. It follows that

$$\begin{aligned}\varphi\{\mu_x(\varphi(x)\neq 3\,\&\,x\geqslant n)\}&=\varphi\{\mu_x(a(n)=0\,\&\,x\geqslant n)\}\\&=\Gamma\{\mu_x(a(n)=0\,\&\,x\geqslant n)\}=q(n),\ \text{say.}\end{aligned}$$

Denoting by $r(n)$ the remainder when n is divided by 10, and by $s(n)$ the quotient $\{n-r(n)\}/10$, we define

$$\psi(n+1)=r\{3\cdot\varphi(n)+s[3\cdot q(n+1)]\}$$

so that $\psi(n+1)$ is formed by multiplying $\varphi(n)$ by 3 and then adding 1 or 0 according as the first $\varphi(k)$, for $k\geqslant n+1$, which is not 3, has the value 5 or 1.

The significance of this definition is best understood by considering the carrying figures when a decimal with digits 1, 3, 5 only is multiplied by 3.

We prove first that the recursive real numbers

$$\left\{\sum_{k\leqslant n}\frac{\psi(k+1)}{10^k}\right\}\ \text{and}\ \left\{3\sum_{k\leqslant n}\frac{\varphi(k)}{10^k}\right\}\ \text{are equal.}$$

Now $\sum_{k\leqslant n}\frac{\psi(k+1)}{10^k}$ is the decimal expansion to n places of the number $3\sum_{k\leqslant n}\frac{\varphi(k)}{10^k}$ for a large enough value of N. Thus

$$0\leqslant 3\sum_{k\leqslant n}\frac{\varphi(k)}{10^k}-\sum_{k\leqslant n}\frac{\psi(k+1)}{10^k}<\frac{1}{10^n},$$

but

$$0\leqslant 3\sum_{k\leqslant n}\frac{\varphi(k)}{10^k}-3\sum_{k\leqslant n}\frac{\varphi(k)}{10^k}<\frac{2}{10^n}$$

and therefore

$$\left|\sum_{k\leqslant n}\frac{\psi(k+1)}{10^k}-3\sum_{k\leqslant n}\frac{\varphi(k)}{10^k}\right|<\frac{2}{10^n}$$

which proves the equality of the real numbers in question. The following table shows the values of $\psi(n+1)$ for all possible combinations of values of $\varphi(n)$ and $q(n+1)$.

$\varphi(n)$	$q(n+1)$	$\psi(n+1)$
1	1	3
3	1	9
5	1	5
1	5	4
3	5	0
5	5	0

Hence if we define $t(n)$ to take the value 1 for odd values of n, and the value 5 for even n, we have

$$q(n+1)=t(\psi(n+1)).$$

If therefore $\psi(n)$ were primitive recursive, then $q(n)$ would be primitive recursive, and by the previous theorem we know that this is not the case.

In classical analysis an expansion like $\Sigma\psi(n+1)\cdot 10^{-n}$ in which $\psi(n)\neq 9$ for infinitely many n, is unique, but the classical proof of this is not valid in recursive analysis. The above example does not therefore answer conclusively the question whether there is a primitive recursive real number which *cannot* have a primitive recursive expansion.

Chapter II

RECURSIVE AND RELATIVE CONTINUITY

Uniform and doubly uniform equivalents of a relatively continuous recursive function. Upper and lower bounds and the impossibility of their attainment. Conditions which ensure that a relatively continuous function vanishes if it changes sign and the non-existence of a recursive root in the general case.

In this chapter we study the properties of recursive continuity and relative continuity, the former a property of single recursive functions and the latter a property of convergent sequences of recursive functions. We define, first, recursive continuity at a point and in an interval.

2. **Recursive continuity**

2, 0. 1 A recursive function $f(x)$ is *recursively continuous at* x_1 if there is a recursive function $c_1(k)$ such that, for all x,

$$x - x_1 = 0(c_1(k)) \rightarrow f(x) - f(x_1) = 0(k).$$

2, 0. 2 $f(x)$ is *uniformly* (*or interval*) *recursively continuous* for $a \leqslant x \leqslant b$ if there is a $c(k)$ such that, for all $x, X, a \leqslant x \leqslant X \leqslant b$,

$$x - X = 0(c(k)) \rightarrow f(x) - f(X) = 0(k).$$

Thus, for instance, the function x^2 is uniformly recursively continuous in any interval $(-10^p, 10^p)$ since $x^2 - X^2 = 0(k)$ if

$$x - X = 0(k + p + 1).$$

If $f(x), g(x)$ are recursively continuous at x_1 (in an interval i) then so of course are $f(x) \pm g(x), f(x) \cdot g(x)$ and also $f(x)/g(x)$, provided that $g(x_1) \neq 0$ ($|g(x)| \geqslant 10^{\alpha}$ in i).

2, 1 Relative continuity

If $f(n, x)$ is recursively convergent for $a \leqslant x \leqslant b$ and if there is a strictly increasing recursive function $c(k)$ and a recursive $C(k, x, y)$

such that for all x, X satisfying $a \leqslant x \leqslant X \leqslant b$ and $x - X = 0(c(k))$, $f(n, x) - f(n, X) = 0(k)$, for all $n \geqslant C(k, x, X)$, then we say that $f(n, x)$ is *continuous for* $a \leqslant x \leqslant b$, *relative to* n. (We observe that relative continuity is a uniform property). A function $f(n, x)$ which is continuous relative to n is not necessary continuous for any particular values of n. Thus for instance if $\varphi(p/q) = Q$, where $Q = q/(p, q)$ and (p, q) is the highest common factor of p and q then $\varphi(x)$ is not continuous for any rational value of x, for if p/q and p_1/q_1 are in their lowest terms with $0 < |p/q - p_1/q_1| < 1/2q_1^2$ then $q > 2q_1$ and so $\varphi(p/q) - \varphi(p_1/q_1) = q - q_1 > q_1 \geqslant 1$; hence if $f(n, x) = \varphi(x)/n$, then $f(n, x)$ is not continuous for any n, but

$$|f(n, p_1/q_1) - f(n, p_2/q_2)| \leqslant |q_1 - q_2|/n = 0(k)$$

for $n > |q_1 - q_2| 10^k$, so that $f(n, x)$ is relatively continuous.

Like continuity, relative continuity is preserved under addition, subtraction and multiplication; furthermore if $g(n, x)$ is relatively continuous for $a \leqslant x \leqslant b$ and if for some integer $\alpha, |g(n, x)| \geqslant 10^{\alpha}$ for $a \leqslant x \leqslant b$ and majorant n, then $1/g(n, x)$ is relatively continuous for $a \leqslant x \leqslant b$.

THEOREM 2, 1 *Relative continuity is an invariant of the equivalence relation.*

For if $g(n, x)$ is equivalent to $f(n, x)$ and if $f(n, x)$ is continuous for $a \leqslant x \leqslant b$, relative to n, then, for all x_1, x_2 satisfying

$$a \leqslant x_1 \leqslant x_2 \leqslant b, \; x_1 - x_2 = 0(c_f(k+1))$$

and n majorant, we have

$$g(n, x_1) - f(n, x_1) = 0(k+1), \; g(n, x_2) - f(n, x_2) = 0(k+1)$$

$$f(n, x_1) - f(n, x_2) = 0(k+1)$$

so that

$$g(n, x_1) - g(n, x_2) = 0(k),$$

which proves that $g(n, x)$ is continuous, relative to n.

THEOREM 2, 1. 1 *If $f(n, x)$ is relatively continuous for $a \leqslant x \leqslant b$, and $|f(n, x)| \geqslant 10^{-\mu}$ for majorant n, then either $f(n, x) \geqslant 10^{-\mu}$ for all x in (a, b), or $f(n, x) \leqslant -10^{-\mu}$ for all x in (a, b), with n majorant.*

For if x_1, x_2 are any two points in (a, b), we may divide (x_1, x_2) into a finite number of parts such that the values of $f(n, x)$ at any two points in the same part differ by less than $10^{-\mu-1}$, for majorant n, and therefore $f(n, x_1)$, $f(n, x_2)$ have the same sign, for majorant n.

THEOREM 2, 2 *A relatively continuous function has a uniformly convergent relatively continuous equivalent.*

For if $f(n, x)$ is relatively continuous for $a \leqslant x \leqslant b$, there are recursive functions $N(k, x)$ and $c(k)$ such that, for all x, X satisfying

$$a \leqslant x \leqslant X \leqslant b,\ x - X = 0(c(k))$$

we have

$$f(n, X) - f(n, x) = 0(k),$$

for majorant n, and

$$N \geqslant n \geqslant N(k, x) \rightarrow f(N, x) - f(n, x) = 0(k).$$

Let $\varphi(k, x) = f(N(k, x), x)$, then

$$p > q \rightarrow \varphi(p, x) - \varphi(q, x) = 0(q)$$

and

$$\begin{aligned}\varphi(k, X) - \varphi(k, x) &= f(N(k, X), X) - f(m, X) \\ &\quad + f(m, X) - f(m, x) + f(m, x) - f(N(k, x), x) \\ &= 3 \cdot 0(k), \text{ for majorant } m,\end{aligned}$$

which shows that $\varphi(n, x)$ is uniformly convergent and relatively continuous.

Since

$$\begin{aligned}\varphi(n, x) - f(n, x) &= \varphi(n, x) - f(m, x) + f(m, x) - f(n, x) \\ &= 2 \cdot 0(k), \text{ for majorant } n,\end{aligned}$$

therefore $\varphi(n, x)$ is equivalent to $f(n, x)$.

We observe further that, for $n \geqslant k$, and $X - x = 0(c(k))$,

$$\varphi(n, X) - \varphi(n, x) = 5 \cdot 0(k)$$

for $\varphi(n, X) - \varphi(k, X) = 0(k)$ and $\varphi(n, x) - \varphi(k, x) = 0(k)$ if $n \geqslant k$, and $\varphi(k, X) - \varphi(k, x) = 3 \cdot 0(k)$, if $X - x = 0(c(k))$.

2, 2 In virtue of Theorem 2, 2 we may, without loss of generality suppose any relatively continuous function $f(n, x)$ to be in standard form, so that $p \geqslant q \rightarrow f(p, x) - f(q, x) = 0(q)$.

THEOREM 2, 2. 1 *If $f(n, x)$ is relatively continuous for $a \leqslant x \leqslant b$ and if $s(n)$ is recursively convergent to a rational x^*, $a < x^* < b$, then $f(n, s(n))$ and $f(n, x^*)$ are equivalent.*

Since $s(n)$ is recursively convergent to x^*, there is a recursive, strictly increasing function $\nu(k)$ such that

$$n \geqslant \nu(k) \rightarrow s(n) - x^* = 0(k)$$

and therefore

$$n \geqslant \nu(c(k)) \rightarrow s(n) - x^* = 0(c(k))$$

whence it follows that

$$n \geqslant \nu(c(k)) \rightarrow f(n, s(n)) - f(n, x^*) = 5 \cdot 0(k)$$

and so $f(n, s(n))$ and $f(n, x^*)$ are equivalent.

In particular if $f(x)$ is recursively continuous for $a \leqslant x \leqslant b$, and $s(n)$ is recursively convergent to $x^*, a < x^* < b$, then $f(s(n))$ is recursively convergent to $f(x^*)$.

2, 2. 1 Whether the sequence $s(n)$ has a rational limit or not, the following form of Theorem 2, 2. 1 holds.

THEOREM 2, 2. 2 *If $f(n, x)$ is continuous for $a \leqslant x \leqslant b$ relative to n and if $s(n)$ is recursively convergent, $a \leqslant s(n) \leqslant b$, then $f(n, s(n))$ is recursively convergent.*

Since $s(n)$ is recursively convergent there is a recursive function $N(k)$ such that

$$N \geqslant n \geqslant N(k) \rightarrow s(N) - s(n) = 0(k),$$

and by relative continuity,

$$\{x - X = 0(c(k))\} \& n \geqslant k \rightarrow f(n, x) - f(n, X) = 5 \cdot 0(k)$$

and therefore, if $N \geqslant n \geqslant N(c(k))$,

$$f(n, s(N)) - f(n, s(n)) = 5 \cdot 0(k).$$

Hence $f(N, s(N)) - f(n, s(n))$

$$= f(N, s(N)) - f(n, s(N)) + f(n, s(N)) - f(n, s(n))$$
$$= 6 \cdot 0(k) \text{ if } N \geqslant n \geqslant N(c(k)), \text{ since } N(c(k)) \geqslant N(k) \geqslant k,$$

which completes the proof.

In particular if $f(x)$ is uniformly recursively continuous for $a \leqslant x \leqslant b$, then $f(s(n))$ is recursively convergent.

THEOREM 2, 2. 2. 1 *If $f(n, x)$ is continuous for $a \leqslant x \leqslant b$, relative to n, and if $s(n), t(n)$ are equivalent recursive real numbers, $a \leqslant s(n) \leqslant b$, $a \leqslant t(n) \leqslant b$, then $f(n, s(n))$ and $f(n, t(n))$ are equivalent recursive real numbers.*

For

$$s(n)-t(n)=0(c(k)), \text{ for majorant } n,$$

and

$$X-x=0(c(k)) \rightarrow f(n, x)-f(n, X)=0(k), \text{ for } n \geqslant k,$$

and so

$$f(n, s(n))-f(n, t(n))=0(k), \text{ for majorant } n.$$

We note again the particular case:
If $f(x)$ is uniformly recursively continuous for $a \leqslant x \leqslant b$ and $s(n), t(n)$ are equivalent recursive real numbers satisfying $a \leqslant s(n) \leqslant b$, $a \leqslant t(n) \leqslant b$, then $f(s(n))$ and $f(t(n))$ are equivalent recursive real numbers.

2, 2. 2 The recursive real number $f(n, s(n))$ may be considered to be the value of the relatively continuous function $f(n, x)$ for the recursive real number argument $s(n)$. Thus the value of a relatively continuous (recursively continuous) function for a recursive real argument is constructed, not defined.

Such a definition as

$$\begin{aligned} f(x) &= 0 \text{ if } x \text{ is rational} \\ &= 1 \text{ if } x \text{ is irrational} \end{aligned}$$

finds no place in recursive analysis. A recursive function is necessarily continuous for a recursive irrational real argument.

2, 2. 3 If $f(n, x)$ is relatively continuous for $a \leqslant x \leqslant b$ and $g(n, t)$ is uniformly convergent for $\alpha \leqslant t \leqslant \beta$, both functions in standard form, and if $a \leqslant g(n, t) \leqslant b$ for $\alpha \leqslant t \leqslant \beta$ then

$$f(k+1, g(c_f(k+1), t))$$

is called the application of f to g or the result of substituting g in f and is denoted by $fg(k, t)$.

THEOREM 2, 2. 3 *For* $n \geqslant k+1$, $m \geqslant c_f(k+1)$ *and* $\alpha \leqslant t \leqslant \beta$,

$$fg(k, t) - f(n, g(m, t)) = 0(k).$$

In fact, since

$$g(m, t) - g(c_f(k+1), t) = 0(c_f(k+1) \text{ for } m \geqslant c_f(k+1)$$

therefore

$$fg(k, t) - f(k+1, g(m, t)) = 3 \cdot 0(k+1)$$

and since, for $n \geqslant k+1$,

$$f(n, g(m, t)) - f(k+1, g(m, t)) = 0(k+1)$$

therefore

$$fg(k, t) - f(n, g(m, t)) = 4 \cdot 0(k+1).$$

THEOREM 2, 2. 4 *$fg(k, t)$ is recursively uniformly convergent in standard form.*

For if $q > p$,

$$\begin{aligned} fg(q, t) - fg(p, t) &= f(q+1, g(c_f(q+1), t)) - f(p+1, g(c_f(q+1), t)) \\ &\quad + f(p+1, g(c_f(q+1), t)) - f(p+1, g(c_f(p+1), t)) \\ &= 0(p+1) + 3 \cdot 0(p+1) = 0(p). \end{aligned}$$

THEOREM 2, 2. 5 *If $\varphi(n, x)$, $\gamma(n, t)$ are equivalents of a relatively continuous $f(n, x)$ and uniformly convergent $g(n, t)$, all in standard form then $\varphi\gamma(k, t)$ is equivalent to $fg(k, t)$.*

Since $g(m, t)$ and $\gamma(m, t)$ are equivalent there is a recursive $M(k, t)$ such that

$$m \geqslant M(k, t) \to g(m, t) - \gamma(m, t) = 0(k),$$

and therefore

$$n \geqslant k \,\&\, m \geqslant M(c_f(k), t) \to f(n, g(m, t)) - f(n, \gamma(m, t)) = 5 \cdot 0(k).$$

But by the equivalence of $f(n, x)$ and $\varphi(n, x)$

$$f(n, \gamma(m, t)) - \varphi(n, \gamma(m, t)) = 0(k), \text{ for majorant } n,$$

and therefore

$$f(n, g(m, t)) - \varphi(n, \gamma(m, t)) = 0(k), \text{ for majorant } m, n,$$

and the result now follows from Theorem 2, 2. 3.

2, 2. 4 If $f(n, x)$ is relatively continuous and $g(n, t)$ is uniformly recursively convergent, neither necessarily in standard form, and if $F(n, x)$, $G(n, t)$ are standard form equivalents then we define $fg(n, t)$ to be $FG(n, t)$.

THEOREM 2, 2. 6 *If $f(n, x)$ is relatively continuous for $a \leqslant x \leqslant b$ and if $g(n, t)$ is relatively continuous for $\alpha \leqslant t \leqslant \beta$, with $a < g(n, t) < b$, for majorant n and $\alpha \leqslant t \leqslant \beta$, then $fg(n, t)$ is relatively continuous for $\alpha \leqslant t \leqslant \beta$.*

We may, without loss of generality suppose that f and g are in standard form since any equivalent G of g also satisfies

$$a < G(n, t) < b, \text{ for majorant } n, \text{ and } \alpha \leqslant t \leqslant \beta.$$

Then, if

$$\alpha \leqslant t \leqslant T \leqslant \beta, \text{ and } T - t = 0(c_g(c_f(k+1)+1))$$

$$fg(n, T) - fg(n, t) = f(n+1, g(c_f(n+1), T)) - \\ - f(n+1, g(c_f(n+1), t)) = 0(k),$$

for $n \geqslant k$.

THEOREM 2, 3 *A relatively continuous function is absolutely bounded.*

Let $f(n, x)$ be relatively continuous in (a, b); divide (a, b) into p parts each of length $0(c(1))$ so that, for any two x, X in the same part, and $n \geqslant 1$, $f(n, x) - f(n, X) = 5 \cdot 0(1)$. It follows that for any x in (a, b) and $n \geqslant 1$,

$$|f(n, x) - f(n, a)| < \tfrac{1}{2}p,$$

and, of course, since $f(n, a)$ is recursively convergent in standard form, $f(n, a) - f(1, a) = 0(1)$, for $n \geqslant 1$.

2, 3 Theorem 2, 3 is a consequence of the uniformity of relative continuity. We shall illustrate this by constructing a recursive function $f(n, x)$ which is recursively convergent, and continuous for each n, but unbounded. Let a_n be the decimal expansion of $\sqrt{2}$ to n decimal places, $a_0 = 1$ and $b_n = a_n + 10^{-n}$, $n \geqslant 0$. Further, let $f(0, x) = 0$ and

$$\begin{aligned} f(r+1, x) &= f(r, x), \text{ if } a_0 \leqslant x \leqslant a_r \text{ or } b_r \leqslant x \leqslant b_0, \\ &= r + (x - a_r)/(a_{r+1} - a_r), \text{ if } a_r < x \leqslant a_{r+1}, \\ &= r + (x - b_r)/(b_{r+1} - b_r), \text{ if } b_{r+1} \leqslant x < b_r, \\ &= r + 1, \text{ if } a_{r+1} \leqslant x \leqslant b_{r+1}. \end{aligned}$$

Let x lie in (a, b); if $x^2 < 2$, let a_{r+1} be the first a for which $x < a_{r+1}$, then $f(n, x)$ is constant for $n \geqslant r$. Similarly if $x^2 > 2$ and b_{r+1} is the first b for which $b_{r+1} < x$, then $f(n, x)$ is constant for $n \geqslant r$. Thus $f(n, x)$ is recursively convergent, and obviously $f(n, x)$ is continuous for each fixed n. But $f(n, x)$ is not absolutely bounded, for $f(r, a_r) = r$, and so $f(n, x)$ is not relatively continuous, as may also be seen by noting that $f(r+1, a_{r+1}) < f(r+1, a_r) = 1$ while $a_{r+1} - a_r$ may be made as small as we please.

THEOREM 2, 4 *If $f(n, x)$ is relatively continuous in (a, b) then there is a recursively convergent $h(n)$ such that, for $a \leqslant x \leqslant b$, $f(n, x) \leqslant h(n)$ relative to n.*

The inequality $f(n, x) \leqslant h(n)$ holds relative to n if

$$f(n, x) \leqslant h(n) + 10^{-k}$$

for majorant n.

Let 10^λ be the smallest power of 10 which exceeds $b-a$,

$$a_r^n = a + (b-a)r/10^{c(n)+\lambda},$$

$$h(n) = \max_{0 \leqslant r \leqslant c(n)+\lambda} f(n, a_r^n),$$

and let $k(n)$ be the least r for which $f(n, a_r^n) = h(n)$, and $a^n = a^n_{k(n)}$. We prove that $h(n)$ is recursively convergent.

Since $a^n = a_p^N$, for some p, if $N \geqslant n$, therefore

$$\begin{aligned} h(N) - h(n) &= f(N, a^N) - f(N, a^n) + f(N, a^n) - f(n, a^n) \\ &\geqslant f(N, a^n) - f(n, a^n) > -10^{-n}, \text{ for } N \geqslant n; \end{aligned}$$

let σ be the greatest integer for which $a_\sigma^n \leqslant a^N$, then $a^N < a^n_{\sigma+1}$ and so

$$\begin{aligned} h(N) - h(n) &= f(N, a^N) - f(n, a^N) + f(n, a^N) - f(n, a_\sigma^n) + f(n, a_\sigma^n) - f(n, a^n) \\ &\leqslant \{f(N, a^N) - f(n, a^N)\} + \{f(n, a^N) - f(n, a_\sigma^n)\} \\ &< 6/10^{k+1}, \text{ for } N \geqslant n \geqslant k+1, \end{aligned}$$

whence it follows that, for $N \geqslant n \geqslant k+1$, $h(N) - h(n) = 0(k)$.

Take any x in (a, b) and let τ be the greatest integer for which $a_\tau^n \leqslant x$, so that x lies between a_τ^n and $a^n_{\tau+1}$ and therefore

$$f(n, x) - f(n, a_\tau^n) = 0(k)$$

for $n \geqslant k+1$. It follows that

$$f(n, x) \leqslant h(n) + 10^{-k}$$

for $n \geqslant k+1$.

In a similar way we may determine the lower bound $l(n)$.

2, 4 If the sequence a^n were recursively convergent then since $f(n, a^n) = h(n)$ we could say that the function $f(n, x)$ attains its maximum, the recursive real number $h(n)$, for the recursive real value a^n of its argument. The following example shows, however, that a proof of the recursive convergence of a^n is impossible.

Let $g(n)$ be any primitive recursive function and let $\gamma(n)$ be the primitive recursive function which takes the value n if $g(r) = 0$ for all $r \leqslant n$, and takes the value p if p is the first value of $r \leqslant n$ for which $g(r) > 0$. Thus $\gamma(n) \leqslant n$ for all n, and $\gamma(n) < n$ for some n only if there is a value of n for which $g(n) > 0$. We shall show that a proof of the recursive convergence of a^n constitutes a decision procedure for the equation

$$g(n) = 0$$

but as we have observed the existence of such a decision procedure would entail a contradiction in recursive arithmetic.

Let

$$\begin{aligned} f(n, x) &= (1 - 2^{-\gamma(n)})x, \text{ if } 0 \leqslant x \leqslant 1, \\ &= -(1 - 2^{1-n})x, \text{ if } -1 \leqslant x \leqslant 0. \end{aligned}$$

Then, if $\gamma(n) = n$, the greatest value of $f(n, x)$ occurs when $x = 1$, but if $\gamma(n) < n$ the greatest value of $f(n, x)$ occurs when $x = -1$. Let a^n be the value of x for which $f(n, x)$ takes its greatest value; if a^n is recursive convergent there is a recursive function n_k such that

$$n \geqslant n_k \rightarrow a^n - a^{n_k} = 0(k)$$

and so for $n \geqslant n_1$, $|a^n - a^{n_1}| < 1/10$, and therefore, since a^n is an integer, $a^n = a^{n_1}$ for all n.

Accordingly, if $a^{n_1} = 1$ then $a^n = 1$ for all $n \geqslant n_1$ and so $\gamma(n) = n$ for all n and $g(n) = 0$ for all n; but if $a^{n_1} = -1$ then $\gamma(n_1) < n_1$ and there is a value of $r \leqslant n_1$ for which $g(n) \neq 0$, showing that a proof of recursive convergence for a^n would constitute a decision procedure for the equation $g(n)=0$. The sequence a^n is of course classically convergent and is an example of a convergent sequence which is not recursively convergent.

2, 4. 1 Although Theorem 2, 2 suffices for all our purposes the following result shows how much that theorem may be strengthened.

THEOREM 2, 5 *Every relatively continuous function has a doubly uniformly continuous equivalent.*

More precisely, if $f(n, x)$ is recursively convergent in n and continuous relative to n for $a \leqslant x \leqslant b$, then there are recursive functions $F(n, x)$, and $c_F(k)$ such that

1. $F(n, x)$ is equivalent to $f(n, x)$ for $a \leqslant x \leqslant b$.
2. $a \leqslant x \leqslant y \leqslant b \,\&\, x-y=0(c_F(k)) \rightarrow F(n, x)-F(n, y)=0(k)$ for all n.
3. $a \leqslant x \leqslant b$ and $m \geqslant n \geqslant k+1 \rightarrow F(m, x)-F(n, x)=0(k)$.

Let $\varphi(n, x)=f(N(n, x), x)$ then

$$m \geqslant n \geqslant k \rightarrow \varphi(m, x)-\varphi(n, x)=0(k)$$

and

$$n \geqslant N(k, x) \rightarrow \varphi(n, x)-f(n, x)=0(k).$$

Denoting the greatest of $N(k, x)$, $N(k, y)$, $C(k, x, y)$ by $M(k, x, y)$, we have

$$\begin{aligned} \varphi(n, x)-\varphi(n, y) &= f(N(n, x), x)-f(M(k, x, y), x)+f(M(k, x, y), x)- \\ &\quad -f(M(k, x, y), y)+f(M(k, x, y), y)-f(N(n, y), y) \\ &= 3\cdot 0(k), \text{ for each } n \geqslant k \text{ and } x-y=0(c_f(k)). \end{aligned}$$

If γ is the least integer, positive or negative, such that 10^γ exceeds $b-a$, and if $l(k)=c_f(k)+\gamma$ we define

$$\Delta_r=(b-a)\, 10^{-l(r)}$$

and

$$a_r^n=a+r\cdot\Delta_n$$

so that $\Delta_n=0(c_f(k))$ for $n \geqslant k$ and since $c_f(k+1)>c_f(k)$, $\Delta_{n+1} \leqslant \Delta_n/10$.

It follows that

$$n \geqslant k \to \varphi(n, a_{r+1}^n) - \varphi(n, a_r^n) = 3 \cdot 0(k).$$

We define next a polygonal approximation to $\varphi(n, x)$; let

$$F(n, x) = \varphi(n, a_r^n) + \{\varphi(n, a_{r+1}^n) - \varphi(n, a_r^n)\}\ (x - a_r^n)/\Delta_n$$

for

$$a_r^n \leqslant x \leqslant a_{r+1}^n \text{ and } 1 \leqslant r+1 \leqslant 10^{l(n)},$$

so that $F(n, x)$ is recursive and $F(n, a_r^n) = \varphi(n, a_r^n)$, and for $n \geqslant k$, and $a_r^n < x < a_{r+1}^n$,

$$F(n, x) - F(n, a_r^n) = 3 \cdot 0(k),\ F(n, x) - F(n, a_{r+1}^n) = 3 \cdot 0(k).$$

Consider now any two x, y in (a, b), $x < y$, satisfying $x - y = 0(c_f(k))$; let $n \geqslant k$ and let μ, ν be the greatest and least values of r such that $x < a_r^n$ and $a_r^n < y$ respectively. Then since $a_\nu^n - a_\mu^n = 0(c_f(k))$ it follows that

$$\begin{aligned} F(n, x) - F(n, y) = F(n, x) - F(n, a_\mu^n) + F(n, a_\mu^n) - \\ - F(n, a_\nu^n) + F(n, a_\nu^n) - F(n, y) \\ = 9 \cdot 0(k). \end{aligned}$$

Consider now values of n less than k; since $c_f(k) \geqslant c_f(n) + 1$ therefore

$$x - y < 1/10^{c_f(k)} \leqslant 1/10^{c_f(n)+1}.$$

But

$$a_{r+1}^n - a_r^n \geqslant 1/10^{c_f(n)+1}$$

and therefore at most one a_r^n lies between x and y. If x and y lie in the same closed subinterval (a_r^n, a_{r+1}^n), since $F(n, x)$ is linear in x in each subinterval

$$\frac{F(n, x) - F(n, y)}{x - y} = \frac{F(n, a_r^n) - F(n, a_{r+1}^n)}{a_r^n - a_{r+1}^n} = \frac{3}{\Delta_n} \cdot 0(n) = \frac{3}{\Delta_k} \cdot 0(k);$$

and if x, y lie in adjacent intervals with common end point a_r^n then

$$\frac{F(n, x) - F(n, a_r^n)}{x - a_r^n} = \frac{3}{\Delta_k} \cdot 0(k),\quad \frac{F(n, a_r^n) - F(n, y)}{a_r^n - y} = \frac{3}{\Delta_k}\, 0(k)$$

whence, since

$$\frac{F(n, x)-F(n, y)}{x-y}$$

lies between $\dfrac{F(n, x)-F(n, a_r{}^n)}{x-a_r{}^n}$ and $\dfrac{F(n, a_r{}^n)-F(n, y)}{a_r{}^n-y}$

therefore

$$\frac{F(n, x)-F(n, y)}{x-y}=\frac{3}{\Delta_k}\,0(k)$$

whether x, y lie in the same subinterval or not, from which it follows that

$$|F(n, x)-F(n, y)|<3\cdot 10^{c_f(k)+\gamma}\cdot 10^{-c_f(k)}\cdot 10^{-k}/(b-a)<3\cdot 10^{-k+1}.$$

Thus, for all values of n,

$$F(n, x)-F(n, y)=0(k)$$

provided that $x-y=0(c_f(k+2))$, which proves (2) with

$$c_F(k)=c_f(k+2).$$

Since, for $a_r{}^n\leqslant x\leqslant a_{r+1}^n$,

$$\begin{aligned}F(n, x)-f(n, x)&=\{F(n, x)-F(n, a_r{}^n)\}+\{\varphi(n, a_r{}^n)-\varphi(n, x)\}+\\&\qquad+\{\varphi(n, x)-f(n, x)\}\\&=7\cdot 0(k)\end{aligned}$$

for $n\geqslant N(k, x)$, therefore $F(n, x)$ is equivalent to $f(n, x)$.

Finally, if $m>n\geqslant k$, for any x in (a, b) we can choose r, s so that

$$a_r{}^n\leqslant a_s{}^m\leqslant x\leqslant a_{s+1}^m\leqslant a_{r+1}^n$$

and therefore

$$\begin{aligned}F(m, x)-F(n, x)&=\{F(m, x)-F(m, a_s{}^m)\}+\{F(n, a_r{}^n)-F(n, x)\}\\&\quad+\{\varphi(m, a_s{}^m)-\varphi(n, a_s{}^m)\}+\{\varphi(n, a_s{}^m)-\varphi(n, a_r{}^n)\}\\&=10\cdot 0(k)\end{aligned}$$

which proves (3).

2, 5 We come now to consider the fundamental property of continuous functions. We shall see that the classical property of continuous functions, that a continuous function vanishes if it changes sign, remains valid for uniform recursive continuity but not for relative continuity. We prove first

THEOREM 2, 6 *If $f(x)$ is uniformly recursively continuous for $a \leqslant x \leqslant b$ and $f(a) \cdot f(b) < 0$ then there is a recursive real number $s(n)$ in (a, b) such that $f(s(n)) = 0$, relative to n.*

We may, without loss of generality suppose that $f(a) < 0$, $f(b) > 0$.

Let $a_0 = a$, $b_0 = b$; if $f\left(\frac{a+b}{2}\right) > 0$ we define $a_1 = a_0$, $b_1 = \frac{a_0+b_0}{2}$ and if $f\left(\frac{a+b}{2}\right) \leqslant 0$ we define $a_1 = \frac{a_0+b_0}{2}$, $b_1 = b_0$. Further if $f\left(\frac{a_n+b_n}{2}\right) > 0$ we define $a_{n+1} = a_n$, $b_{n+1} = \frac{a_n+b_n}{2}$, and if $f\left(\frac{a_n+b_n}{2}\right) \leqslant 0$ then

$$a_{n+1} = \frac{a_n+b_n}{2},\ b_{n+1} = b_n$$

so that a_n and b_n are recursively convergent, and since $b_n - a_n \to 0$, a_n and b_n are equivalent recursive real numbers. It is readily seen that $f(a_n) \leqslant 0$ and $f(b_n) > 0$, for this is true with $n = 0$, and if it is true for $n = k$, then since $f(a_{k+1}) = f(a_k)$ and $f(b_{k+1}) = f\left(\frac{a_k+b_k}{2}\right)$ if $f\left(\frac{a_k+b_k}{2}\right) > 0$; and $f(a_{k+1}) = f\left(\frac{a_k+b_k}{2}\right)$, $f(b_{k+1}) = f(b_k)$ if $f\left(\frac{a_k+b_k}{2}\right) \leqslant 0$, it holds also for $n = k+1$, and therefore for all n.

Since $f(x)$ is uniformly recursively continuous, $f(a_n)$ and $f(b_n)$ are equivalent, that is

$$|f(b_n) - f(a_n)| < 1/10^k, \text{ for majorant } n,$$

and so

$$f(a_n) = f(b_n) = 0, \text{ relative to } n.$$

2, 5. 1. A similar argument may be applied to a relatively continuous function provided that the function satisfies some additional restriction. We prove

THEOREM 2, 6. 1 *If $f(n, x)$ is continuous for $a \leqslant x \leqslant b$, relative to n, and $f(n, a) < 0$, $f(n, b) > 0$, for all n, and if there are recursive functions g, h such that, for $a \leqslant x \leqslant b$, $n \geqslant h(x) \rightarrow |f(n, x)| \geqslant 10^{-g(x)}$ then there is a recursive real number s_n such that $f(n, s_n) = 0$, relative to n.*

Let $m(x) = \max\ (g(x), h(x))$ and $F(x) = f(m(x), x)$. As in Theorem 2, 6 there are equivalent recursive real numbers a_n, b_n such that

$$F(a_n) \leqslant 0,\ F(b_n) > 0,$$

that is,

$$f(m(a_n), a_n) \leqslant 0, \text{ so that } f(m(a_n), a_n) \leqslant -1/10^{g(a_n)}$$

and similarly,

$$f(m(b_n), a_n) \geqslant 1/10^{g(b_n)}.$$

Since

$$|f(p, x) - f(m(x), x)| < 1/10^{m(x)}, \text{ for } p \geqslant m(x),$$

therefore

$$f(p, a_n) < 0 \text{ for } p \geqslant m(a_n),$$

and similarly

$$f(p, b_n) > 0 \text{ for } p \geqslant m(b_n).$$

However, since $b_n = a_n$, relative to n,

$$f(k, b_n) - f(k, a_n) = 3 \cdot 0(k), \text{ for majorant } n,$$

and so

$$f(k, b_n) = 3 \cdot 0(k),\ f(k, a_n) = 3 \cdot 0(k), \text{ for majorant } n,$$

whence it follows that $f(n, a_n) = 0$, relative to n.

2, 6 We consider next the general case of a relatively continuous function, without additional restriction, and prove

THEOREM 2, 6. 2 *If $f(n, x)$ is continuous for $a \leqslant x \leqslant b$, relative to n, then there are recursive functions a_n, b_n such that $a_n \leqslant a_{n+1} < b_{n+1} \leqslant b_n$ and for all x, $a_k \leqslant x \leqslant b_k \rightarrow f(k+2, x) = 0(k)$.*

For any rational x let $[x]$ denote the whole part of x and let $\{x\}_k = [10^k x]/10^k$. Further let

$$a_r^k = a + (b-a)r/10^{c(k+1)+\gamma}$$

(where 10^γ is the smallest power of 10 which exceeds $b-a$). Then for $1 \leqslant r+1 \leqslant 10^{c(k+1)+\gamma}$,

$$f(k+1, a^k_{r+1}) - f(k+1, a_r{}^k) = 3 \cdot 0(k+1)$$

and therefore

$$|\{f(k+1, a^k_{r+1})\}_k - \{f(k+1, a_r{}^k)\}_k| \leqslant 1/10^k,$$

which shows that the integers $10^k\{f(k+1, a_r{}^k)\}_k$ are equal or consecutive for consecutive values of r. *Thus as r varies from*

$$0 \text{ to } 10^{c(k+1)}+\gamma, \{f(k+1, a_r{}^k)\}_k$$

takes every value $n/10^k$ between any two of its values, and this is the essential property which a relatively continuous function has in common with a recursively continuous function. In particular, from $f(n, a) < -10^{-\mu}$ follows $f(k+1, a) < 10^{-k+1}$ for $k \geqslant \mu$, and so $\{f(k+1, a)\}_k \leqslant -10^{-k+1} < -10^{-k}$; similarly from $f(n, b) > 10^{-\mu}$ follows $\{f(k+1, b)\}_k > 10^{-k}$, for $k \geqslant \mu$, and therefore there is a least ν, $\nu = \nu_k$ say, such that $\{f(k+1, a^k_{\nu_k})\}_k = 0$, that is, $f(k+1, a^k_{\nu_k}) = 0(k)$.

We shall subsequently show by an example that in general the sequence $a^k_{\nu_k}$ is *not* recursively convergent. First however we shall analyse the position a little more exactly by showing that there are recursive functions a_n, b_n such that $a_n \leqslant a_{n+1} < b_{n+1} \leqslant b_n$ and

$$f(k+2, x) = 0(k),$$

for all x, $a_k \leqslant x \leqslant b_k$.

We recall first that from $f(k+1, x) < -1/10^k$ follows

$$f(k+2, x) < -1/10^{k+1}, \text{ since } f(k+2, x) - f(k+1, x) = 0(k+1).$$

Let $a_r{}^k = a + (b-a)r/10^{c(k)+\gamma}$, then since $f(\mu+1, a) \leqslant -1/10^\mu$ and $f(\mu+1, b) \geqslant 1/10^\mu$ there is a first r, say $r = r_\mu$ such that

$$f(\mu+1, a^\mu_{r+1}) \geqslant 1/10^\mu,$$

and a greatest $s+1 < r_\mu$, say $s+1 = s_\mu$, such that

$$f(\mu+1, a_s{}^\mu) \leqslant -1/10^\mu,$$

so that

$$-1/10^\mu < f(\mu+1, a_n{}^\mu) < 1/10^\mu$$

for all n from s_μ to r_μ inclusive. We now define r_n, s_n, for $n \geqslant \mu$, recursively as follows: suppose that, for some k,

$$-1/10^k < f(k+1, a_n{}^k) < 1/10^k$$

for all n from s_k to r_k, and that

$$f(k+1, a^k_{s_k}) \leqslant -1/10^k,\ f(k+1, a^k_{r_k}) \geqslant 1/10^k.$$

Then

$$f(k+2, a^k_{s_k}) < -1/10^{k+1},\ f(k+2, a^k_{r_k}) > 1/10^{k+1}.$$

Let r_{k+1} be the first r greater than $10^{c(k+1)-c(k)} \cdot s_k$ such that

$$f(k+2, a^{k+1}_{r+1}) \geqslant 1/10^{k+1}$$

and let s_{k+1} be the greatest $s+1 < r_{k+1}$ such that

$$f(k+2, a_s{}^{k+1}) \leqslant -1/10^{k+1},$$

so that

$$-1/10^{k+1} < f(k+2, a_n{}^{k+1}) < 1/10^{k+1}$$

for all n from s_{k+1} to r_{k+1} inclusive, which completes the definition. Let $a^k_{s_k} = a_k$, $a^k_{r_k} = b_k$. By definition of r_k, s_k, we have, for all $k \geqslant \mu$,

$$-1/10^k < f(k+1, a_n{}^k) < 1/10^k$$

for all n from s_k to r_k inclusive.

If $a_n{}^k \leqslant x \leqslant a_{n+1}$ then $f(k+1, x) - f(k+1, a_n{}^k) = 5 \cdot 0(k)$ so that

$$f(k+1, x) = 6 \cdot O(k).$$

Thus for all x, $a_k \leqslant x \leqslant b_k$, we have $f(k+1, x) = 6 \cdot 0(k)$, and

$$a_k \leqslant a_{k+1} < b_{k+1} \leqslant b_k.$$

Although the sequence a_k is non-decreasing and bounded above, and b_k is non-increasing and bounded below, nevertheless, as we shall see, neither a_k nor b_k is necessarily recursively convergent, and it is not possible to decide in general whether $b_k = a_k$ relative to k, or not. However if $f(n, x)$ satisfies some suitable additional condition we can prove that a_k and b_k are equivalent recursive real numbers. Specifically we prove

THEOREM 2, 6. 3 *If $f(n, x)$ is continuous for $a \leqslant x \leqslant b$, relative to n, and if there is a constant q such that, for $a \leqslant x < y \leqslant b$, $\left|\frac{f(n, x)-f(n, y)}{x-y}\right| > q$, for majorant n, then there is a recursively convergent sequence a_n such that $f(n, a_n)=0$, relative to n.*

By the previous theorem, we can find recursive sequences a_n, b_n such that,

$$a_k \leqslant a_{k+1} < b_{k+1} \leqslant b_k$$

and

$$a_k \leqslant x \leqslant b_k \text{ and } n \geqslant k+2 \rightarrow f(n, x)=0(k).$$

But, by hypothesis

$$0 < b_k - a_k < (1/q)|f(n, b_k)-f(n, a_k)|, \text{ for majorant } n,$$

and therefore

$$0 < b_k - a_k < (2/q)\cdot 0(k)$$

which proves that a_k and b_k are recursively convergent.

2, 6. 1 We show now that the sequences a_k, b_k introduced in Theorem 2, 6. 2 are not necessarily recursively convergent. As in § 2. 4 let $g(n)$ be any primitive recursive function and $\gamma(n)$ the recursive function which takes the value n if $g(r)=0$ for all $r \leqslant n$ and takes the value p if p is the first $r \leqslant n$ for which $g(r)>0$, so that $\gamma(n)=n$ unless there is a value p for which $g(p)>0$, when $\gamma(n)<n$ for $n \geqslant p$. We define $f(n, x)$ in the range $0 \leqslant x \leqslant 3$ as follows:

If $\gamma(n)=n$,

$$\begin{aligned} f(n, x) &= x-1, \text{ for } 0 \leqslant x \leqslant 1, \\ &= 0, \text{ for } 1 \leqslant x \leqslant 2, \\ &= x-2, \text{ for } 2 \leqslant x \leqslant 3; \end{aligned}$$

if $\gamma(n)=p<n$,

$$\begin{aligned} f(n, x) &= x-1, \text{ for } 0 \leqslant x \leqslant 1-1/(p+1), \\ &= -1/(p+1), \text{ for } 1-1/(p+1) \leqslant x \leqslant 2-1/(p+1), \\ &= x-2, \text{ for } 2-1/(p+1) \leqslant x \leqslant 3. \end{aligned}$$

The function $f(n, x)$ so defined is obviously recursively convergent and continuous for $0 \leqslant x \leqslant 3$, relative to n. For if $N \geqslant n$ then

$$\gamma(N)=N \text{ or } \gamma(N)=p \leqslant n \rightarrow f(N, x)-f(n, x)=0$$

and

$$n<\gamma(N)<N \rightarrow 0 \leqslant f(n, x)-f(N, x) \leqslant 1/(n+1);$$

moreover

$$X-x=0(k) \rightarrow f(n, X)-f(n, x)=0(k)$$

for all values of n.

Let a_k denote the first point, and b_k the last, between 0 and 3 where $f(k, x)$ vanishes. Then, if $g(n)=0$ for all n, $a_n=1$ but if p is the first n for which $g(n)>0$ then $a_n=2$, $n \geqslant p$. Accordingly if a_n were recursively convergent we should have a decision procedure for the equation $g(n)=0$, for we should have an integer n_1 such that $a(n)-a(n_1)=0(1)$ for all $n \geqslant n_1$ and therefore $a(n_1) \leqslant 1\frac{1}{2}$ would prove $g(n)=0$ for all n and $a(n_1)>1\frac{1}{2}$ would prove that there is an n for which $g(n)>0$.

We have at the same time shown that it is impossible to decide whether $f(n, x)=0$ relative to n for a single value of x or for an interval of values, since if we could decide this we should again have a decision procedure for the equation $g(n)=0$.

Chapter III

RECURSIVE AND RELATIVE DIFFERENTIABILITY

The mean value inequalities. Doubly uniform equivalent of a relatively differentiable function. The mean value theorem. Taylor's theorem. The uniform mean value theorem. The existence of relatively differentiable functions not satisfying the uniform mean value theorem.

3. Differentiability

3, 0. 1 A recursive function $f(x)$ is said to be *recursively differentiable* at x_0 if there is a recursive $d(k)$ and a number A such that

$$x-x_0=0(d(k)) \rightarrow f(x)-f(x_0)=(x-x_0)(A+0(k)).$$

3, 0. 2 $f(x)$ is *uniformly recursively differentiable* for $a \leqslant x \leqslant b$ if there are recursive functions $f^1(x)$, $d(k)$ such that

$$a \leqslant x < X \leqslant b \;\&\; x-X=0(d(k)) \rightarrow f(x)-f(X)=(x-X)(f^1(x)+0(k)).$$

$f^1(x)$ is called the recursive derivative in (a, b). In general the functions $d(k)$, $f^1(x)$ will contain any parameter in $f(x)$.

3, 0. 3 A recursively convergent function $f(n, x)$ of an integral variable n and rational variable x is said to be *differentiable for* $a \leqslant x \leqslant b$, *relative to* n if there are recursive functions $f^1(n, x)$, $d(k)$ such that, for majorant n,

$$\begin{aligned} a \leqslant x < X \leqslant b \;\&\; x-X=0(d(k)) \rightarrow f(n, x)-f(n, X) \\ =(x-X)(f^1(n, x)+0(k)). \end{aligned}$$

$f^1(n, x)$ is called a relative derivative of $f(n, x)$ in (a, b).

Theorem 3, 0. 1 *If $f(n, x)$ is relatively differentiable in (a, b) with relative derivative $f^1(n, x)$ and if $\varphi(n, x)$, $\varphi^1(n, x)$ are equivalent to $f(n, x)$, $f^1(n, x)$ respectively, then $\varphi(n, x)$ is relatively differentiable in (a, b) with relative derivative $\varphi^1(n, x)$.*

For, if $x<X$,

$$\frac{f(n,x)-f(n,X)}{x-X}-\frac{\varphi(n,x)-\varphi(n,X)}{x-X}=$$
$$=\frac{f(n,x)-\varphi(n,x)}{x-X}+\frac{\varphi(n,X)-f(n,X)}{x-X}$$
$$=0(k),\ \text{for majorant } n,$$

and

$$f^1(n,x)=\varphi^1(n,x)+0(k),\ \text{for majorant } n,$$

and therefore

$$a\leqslant x<X\leqslant b\ \&\ x-X=0(d(k))\rightarrow\varphi(n,x)-\varphi(n,X)=$$
$$=(x-X)(\varphi^1(n,x)+3\cdot 0(k))\ \text{for majorant } n.$$

We note that, in particular, $\varphi^1(n,x)$ is a relative derivative of $f(n,x)$ and $f^1(n,x)$ is a relative derivative of $\varphi(n,x)$.

THEOREM 3, 0. 2 *If $f(n,x)$ is relatively differentiable in (a,b) with relative derivative $f^1(n,x)$ then $f(n,x)$ and $f^1(n,x)$ are continuous for $a\leqslant x\leqslant b$, relative to n.*

From

$$f(n,x)-f(n,X)=(x-X)\{f^1(n,x)+0(k)\},$$

for $x-X=0(d(k))$ and majorant n, follows, interchanging x, X

$$f(n,X)-f(n,x)=(X-x)\{f^1(n,X)+0(k)\},\ \text{for majorant } n,$$

whence

$$f^1(n,X)-f^1(n,x)=2\cdot 0(k),\ \text{for majorant } n,$$

proving that $f^1(n,x)$ is relatively continuous.

Since $f^1(n,x)$ is relatively continuous, $|f^1(n,x)|$ is bounded, by 10^{λ} say.

Let

$$c(k)=\max\{d(k),\ k+\lambda+1\}$$

then

$$X-x=0(c(k))\rightarrow f(n,x)-f(n,X)=0(k),\ \text{for majorant } n,$$

which proves that $f(n,x)$ is relatively continuous.

3, 1 Recursive differentiability for all rational x in some interval does not of course entail uniform recursive differentiability. For instance if $f(x)$ is defined for $x \geqslant 0$ by the conditions

$$x^2 < 2 \rightarrow f(x) = -x$$
$$x^2 > 2 \rightarrow f(x) = +x$$

then $f(x)$ is recursive and recursively differentiable for each fixed rational $x \geqslant 0$. But $f(x)$ is not uniformly recursively differentiable for $1 \leqslant x \leqslant 2$, for if $x^2 < 2$ and $X^2 > 2$ then

$$\frac{f(X)-f(x)}{X-x} = \frac{X+x}{X-x} \geqslant \frac{2}{X-x} > 2N \text{ if } X-x < 1/N$$

so that $\dfrac{f(X)-f(x)}{X-x}$ is unbounded.

Moreover a function $f(n, x)$ may be relatively differentiable in some interval without being uniformly differentiable for any fixed value of n. For suppose that

$$\begin{aligned} f(n, p/q) &= q, \text{ if } q \geqslant n \\ &= 0, \text{ if } q < n; \end{aligned}$$

then $f(n, p/q)$ is recursively convergent since $f(N, p/q) - f(n, p/q) = 0$ for $N \geqslant n > q$, and $f(n, p/q)$ is relatively differentiable in any interval since

$$\frac{f(n, p/q) - f(n, p'/q')}{p/q - p'/q'} = 0 \text{ for } n > \max(q, q').$$

But for any fixed n, and $q > n$

$$\frac{f(n, 1/q) - f(n, 1/(q+1))}{1/q - 1/(q+1)} = -\,q(q+1)$$

which is unbounded, showing that $f(n, x)$ is not uniformly recursively differentiable for fixed n.

THEOREM 3, 1 *If $f(x)$ is recursively differentiable with derivative $f^1(x)$ for $a \leqslant x \leqslant b$, and if $f^1(x) = 0$, then, for $a \leqslant x \leqslant b$, $f(x) = f(a)$.*

Let the points a_r^k, $0 \leqslant r \leqslant i(k)$ divide (a, b) into equal parts of length $\Delta_k = 0(d(k))$, so that, for $a_r^k \leqslant x \leqslant a_{r+1}^k$, $1 \leqslant r+1 \leqslant i(k)$,

$$\{f(x) - f(a_r^k)\}/(x - a_r^k) = 2 \cdot 0(k), \text{ for majorant } n,$$

whence

$$f(x) - f(a) = 0$$

for any x in (a, b).

THEOREM 3, 1. 1 *If $f(n, x)$ is differentiable, with relative derivative $f^1(n, x)$, for $a \leqslant x \leqslant b$, and if $g(n, t)$ is differentiable with relative derivative $g^1(n, t)$ for $\alpha \leqslant t \leqslant \beta$, and if further $a < g(n, t) < b$ for $\alpha \leqslant t \leqslant \beta$, then $fg(n, t)$ is relatively differentiable for $\alpha \leqslant t \leqslant \beta$ with relative derivative $f^1g(n, t) \cdot g^1(n, t)$.*

Consider any two t, T such that $\alpha \leqslant t < T \leqslant \beta$; let $e(t, T)$ be the exponent of the least power of 10 which exceeds $1/(T-t)$, and let

$$d(k) = \max\,(d_g(k), c_g(d_f(k))).$$

Then for $T - t = 0(d(k))$,

$$\frac{f(n, g(m, T)) - f(n, g(m, t))}{T-t} - f^1(n, g(m, t)) \cdot g^1(m, t) =$$
$$= \{f^1(n, g(m, t)) + g^1(m, t) + 0(k)\} \cdot 0(k) \text{ for majorant } m, n.$$

But, for $p > k + e(t, T)$,

$$\frac{fg(p, T) - fg(p, t)}{T-t} = \frac{f(n, g(m, T)) - f(n, g(m, t))}{T-t} + 0(k)$$

for majorant m, n, and

$$f^1g(p, t) \cdot g^1(p, t) - f^1(n, g(m, t)) \cdot g^1(m, t) =$$
$$= \{f^1(n, g(m, t)) + g^1(m, t) + 0(k)\} \cdot 0(k), \text{ for majorant } m, n,$$

whence Theorem 3, 1. 1 follows since $f^1(n, x)$ and $g^1(m, t)$ are bounded.

THEOREM 3, 2 *If $f(n, x)$ is differentiable in (a, b), relative to n, with relative derivative $f^1(n, x)$, and if $f^1(n, x) = 0$, relative to n, for all x in (a, b) then $f(n, x) = f(n, a)$, relative to n, for all x in (a, b).*

Let $a_r^k = a + (b-a)r/10^{d(k)+\gamma}$, where γ is the index of the smallest power of 10 which exceeds $b-a$, so that

$$a_{r+1}^k - a_r^k = 0(d(k)).$$

For any x in (a, b), let $a_s^k < x \leqslant a_{s+1}^k$, then

$$\begin{aligned}\frac{f(n, a_{s+1}^k) - f(n, x)}{a_{s+1}^k - x} &= f^1(n, x) + 0(k), \text{ for majorant } n,\\ &= 2 \cdot 0(k), \text{ for majorant } n,\end{aligned}$$

and for each r, $0 \leqslant r < s$,

$$\begin{aligned}\frac{f(n, a_{r+1}^k) - f(n, a_r^k)}{a_{r+1}^k - a_r^k} &= f^1(n, a_r^k) + 0(k), \text{ for majorant } n,\\ &= 2 \cdot 0(k), \text{ for majorant } n,\end{aligned}$$

whence

$$\frac{f(n, x) - f(n, a)}{x - a} = 2 \cdot 0(k) \text{ for majorant } n$$

and so $f(n, x) = f(n, a)$, relative to n.

3, 2 The mean value inequalities

Theorem 3, 2. 1 *If, for each value of n, $f(n, x)$ is uniformly recursively differentiable for $a \leqslant x \leqslant b$, with recursive derivatives $f^1(n, x)$ then there is a recursive c_k^n such that $a < c_k^n < b$ and*

$$\frac{f(n, b) - f(n, a)}{b - a} \leqslant f^1(n, c_k^n) + 0(k)$$

for each value of n.

Let $\mu_n(x, y)$ denote $\{f(n, x) - f(n, y)\}/(x-y)$, $x < y$; if z is the mid point of (x, y) then $\mu_n(x, y)$ lies between $\mu_n(x, z)$ and $\mu_n(z, y)$ and so $\mu_n(x, y)$ is exceeded by one of $\mu_n(x, z)$, $\mu_n(z, y)$; thus we may bisect (a, b) repeatedly, choosing a succession of intervals (a, b), (a_1^n, b_1^n), (a_2^n, b_2^n), ..., say, each of which is half its predecessor, and such that $\mu_n(a_r^n, b_r^n)$ is non-decreasing in r.

For a suitable value of r, however,

$$\mu_n(a_r^n, b_r^n) = f^1(n, c_k^n) + 0(k),$$

where we take $c_k{}^n$ to be either $a_r{}^n$, or $b_r{}^n$, whichever is in the interior of (a, b).

The same argument also proves

THEOREM 3, 2. 2 *Under the conditions of Theorem* 3, 2. 1 *we can find a recursive* $c_k{}^n$ *in* (a, b) *such that*

$$\{f(n, b) - f(n, a)\}/(b-a) \geqslant f^1(n, c_k{}^n) + 0(k),$$

for each value of n.

3, 2. 1 The mean value inequalities hold also for relative differentiability. For if $\mu_n(a_r{}^n, b_r{}^n)$ is determined as above, and if $f(n, x)$ is differentiable for $a \leqslant x \leqslant b$, relative to n, then for an r satisfying $(b-a)/2^r = 0(d(k))$ we have

$$\mu_n(a_r{}^n, b_r{}^n) = f^1(n, c_k{}^n) + 0(k)$$

for majorant n, where $c_k{}^n$ is either $a_r{}^n$ or $b_r{}^n$. Thus we have

THEOREM 3, 2. 3 *If $f(n, x)$ is relatively differentiable for $a \leqslant x \leqslant b$, with relative derivative $f^1(n, x)$, then there is a recursive $c_k{}^n$ in (a,b) such that*

$$\{f(n, b) - f(n, a)\}/(b-a) \leqslant f^1(n, c_k{}^n) + 0(k)$$

for majorant n,

and the corresponding result with the inequality reversed.

3, 3 We can however prove a far stronger result than Theorem 3,2.3 as we shall subsequently show (Theorem 3, 3 below).

THEOREM 3, 2. 4 *If, for $a \leqslant x \leqslant b$, $s(n, x)$ is uniformly recursively differentiable for each fixed n, with derivative $\sigma(n, x)$, and if $\sigma(n, x)$ is uniformly recursively convergent, then $s(n, x)$ is differentiable relative to n, with relative derivative $\sigma(n, x)$.*

It follows from definition 3, 0. 2 that there is a recursive $d(n, k)$ such that

$$a \leqslant x < y \leqslant b \,\&\, x - y = 0(d(n, k)) \rightarrow \sigma(n, x) - \sigma(n, y) = 0(k),$$

and since $\sigma(n, x)$ is uniformly recursively convergent there is an $N(k)$ such that

$$\sigma(n, x) - \sigma(N(k), x) = 0(k), \text{ for } n \geqslant N(k).$$

By Theorem 3, 2. 1 there is a $c_k{}^n$ such that $a \leqslant x < c_k{}^n < y \leqslant b$ and

$$\frac{s(n, x) - s(n, y)}{x - y} \leqslant \sigma(n, c_k{}^n) + 0(k),$$

and therefore, for $n \geqslant N(k)$ and $x - y = 0(d(N(k), k))$,

$$\begin{aligned}\frac{s(n, x) - s(n, y)}{x - y} - \sigma(n, x) \leqslant \{\sigma(n, c_k{}^n) - \sigma(N(k), c_k{}^n)\} \\ + \{\sigma(N(k), c_k{}^n) - \sigma(N(k), x)\} \\ + \{\sigma(N(k), x) - \sigma(n, x)\} + 0(k) = 4 \cdot 0(k),\end{aligned}$$

and similarly, by Theorem 3, 2. 2, for the same n, x, y,

$$\frac{s(n, x) - s(n, y)}{x - y} - \sigma(n, x) \geqslant 4 \cdot 0(k),$$

which completes the proof.

Theorem 3, 2. 5 *If*

(i) *for* $a \leqslant x \leqslant b$, $g(m, x)$ *is continuous relative to* m, *and*

$$\alpha = g(m, a) \leqslant g(m, x) \leqslant \beta, \text{ for majorant } m,$$

(ii) *for* $\alpha \leqslant t \leqslant \beta$, $f(n, t)$ *is differentiable relative to* n *and*

$$|f^1(n, t)| \geqslant 10^{-\mu} \text{ for majorant } n, \text{ and } f(n, \alpha) = a,$$

(iii) *for* $a \leqslant x \leqslant b$, $fg(n, x) = x$, *relative to* n,

then

$$gf(n, t) = t, \text{ relative to } n,$$

for any t *such that, for an* $x < b$, $\alpha \leqslant t < g(m, x)$ *for majorant* m.

It follows from condition (ii) that $f^1(n, t)$ is of constant sign, for $\alpha \leqslant t \leqslant \beta$ and majorant n, and so we may, without loss of generality, suppose that $f^1(n, t) \geqslant 10^{-\mu}$ for $\alpha \leqslant t \leqslant \beta$ and majorant n. From (iii) it follows that $f(n, g(m, x)) = x + 0(k)$, for majorant m, n.

We commence by proving that, if $\alpha \leqslant t < g(m, x)$ for majorant m and some $x < b$, then $a \leqslant f(n, t) < b$, for majorant n.

For the inequality which $f^1(n, t)$ satisfies shows, by Theorem 3, 2. 3

that $f(n, t)$ is strictly increasing (for majorant n), and so by (iii) if $\alpha \leqslant t < g(m, x)$, for majorant m, and $x < b$, then

$$a = f(n, \alpha) \leqslant f(n, t) < f(n, g(m, x)) = x + 0(k) < b$$

for a large enough k, and majorant m, n.

Using the uniform convergence of $f(n, x)$, $g(n, x)$ and the relative continuity of $f(n, x)$, it follows from (iii) that

$$f(n, g(m, x)) = x + 5 \cdot 0(k)$$

for $m \geqslant c_f(k)$, $n \geqslant k$ and $a \leqslant x \leqslant b$, and therefore, for the same m, n and for $p \geqslant k$

$$f(n, g(m, f(p, t)) = f(n, t) + 6 \cdot 0(k)$$

for $\alpha \leqslant t < g(m, x)$ and $x < b$.

However, by Theorem 3, 2. 2, there is a recursive $c_k{}^n$ such that, writing g for short for $g(m, f(p, t))$,

$$|f(n, g) - f(n, t)| \geqslant |g - t| \cdot \{f^1(n, c_k{}^n) + 0(k)\}$$

for majorant n, and therefore,

$$g(m, f(p, t)) - t = 0(k - \mu - 1)$$

for $m \geqslant c_f(k)$, $p \geqslant k$, whence it follows that

$$gf(n, t) = t, \text{ relative to } n.$$

Theorem 3, 3 The Mean Value Theorem. *If $f(n, x)$ is differentiable for $a \leqslant x \leqslant b$, relative to n, then we can determine c_k, depending only upon k, such that $a < c_k < b$ and*

$$\frac{f(n, b) - f(n, a)}{b - a} = f^1(n, c_k) + 0(k)$$

for majorant n.

Let $a_r{}^n = a + (b - a)r/10^{c_{f'}(n) + \gamma}$, where $(b - a) \leqslant 10^\gamma$, and

$$\Delta_n = a_{r+1}^n - a_r{}^n,$$

and define

$$g(n, x) = f^1(n, a_r{}^n) + \frac{x - a_r{}^n}{\Delta_n} \{f^1(n, a_{r+1}^n) - f^1(n, a_r{}^n)\}$$

for $a_r{}^n \leqslant x \leqslant a_{r+1}^n$, $0 \leqslant r < t_n = 10^{c_{f^1}(n)+\gamma}$, and

$$G(n, a) = f(n, a),$$

$$G(n, x) = G(n, a_r{}^n) + (x - a_r{}^n) \cdot f^1(n, a_r{}^n) + \\ + \frac{(x - a_r{}^n)^2}{2\Delta_n} \{f^1(n, a_{r+1}^n) - f^1(n, a_r{}^n)\}$$

for $a_r{}^n < x \leqslant a_{r+1}^n$, $0 \leqslant r < t_n$.

Since

$$g(n, x) - f^1(n, x) = \{f^1(n, a_r{}^n) - f^1(n, x)\} + \frac{x - a_r{}^n}{\Delta_n} \{f^1(n, a_{r+1}^n) - f^1(n, a_r{}^n)\}$$
$$= 6 \cdot 0(k) \text{ for } n \geqslant k \text{ and } a \leqslant x \leqslant b,$$

therefore $g(n, x)$ is equivalent to $f^1(n, x)$, so that $g(n, x)$ is relatively continuous.

More precisely, if $a_r{}^n \leqslant x < y \leqslant a_{r+1}^n$

$$g(n, x) - g(n, y) = \frac{(x - y)}{\Delta_n} \{f^1(n, a_{r+1}^n) - f^1(n, a_r{}^n)\} = 5 \cdot 0(k) \text{ for } n \geqslant k,$$

and if

$$x < a_r{}^n < y,\ y - x = 0(c_{f^1}(k)),$$

then

$$g(n, y) - g(n, a_r{}^n) = \frac{(y - a_r{}^n)}{\Delta_n} \{f^1(n, a_{r+1}^n) - f^1(n, a_r{}^n)\} = 5 \cdot 0(k), \text{ for } n \geqslant k,$$

and

$$g(n, x) - g(n, a_r{}^n) = \frac{(a_r{}^n - x)}{\Delta_n} \{f^1(n, a_r{}^n) - f^1(n, a_{r+1}^n)\} = 5 \cdot 0(k), \text{ for } n \geqslant k,$$

so that $g(n, y) - g(n, x) = 0(k-1)$, for $n \geqslant k$ in both cases.

Furthermore, if $a_r{}^n \leqslant x < a_{r+1}^n$,

$$g(N, x) - g(n, x) = \{g(N, x) - g(N, a_r{}^n)\} + \{g(N, a_r{}^n) - g(n, a_r{}^n)\} + \\ + \{g(n, a_r{}^n) - g(n, x)\}$$
$$= 2 \cdot 0(k-1) + \{f^1(N, a_r{}^n) - f^1(n, a_r{}^n)\},\ N \geqslant n \geqslant k,$$
$$= 3 \cdot 0(k-1) \text{ for } N \geqslant n \geqslant k.$$

proving that $g(n, x)$ is uniformly convergent.

We prove next that $G(n, x)$ is differentiable for $a \leqslant x \leqslant b$, relative to n, with relative derivative $g(n, x)$.

If $a_{r-1}^n < x \leqslant a_r^n < y < a_{r+1}^n$ then

$$\frac{G(n, y)-G(n, a_r^n)}{y-a_r^n} = f^1(n, a_r^n) + \frac{y-a_r^n}{2\Delta_n}\{f^1(n, a_{r+1}^n)-f^1(n, a_r^n)\}$$

and

$$\frac{G(n, a_r^n)-G(n, x)}{a_r^n-x} = f^1(n, a_{r-1}^n) +$$

$$+\frac{x+a_r^n-2a_{r-1}^n}{2\Delta_n}\{f^1(n, a_r^n)-f^1(n, a_{r-1}^n)\}$$

$$= f^1(n, a_r^n) + \frac{x-a_r^n}{2\Delta_n}\{f^1(n, a_r^n)-f^1(n, a_{r-1}^n)\}$$

and since

$\dfrac{G(n, y)-G(n, x)}{y-x}$ lies between $\dfrac{G(n, y)-G(n, a_r^n)}{y-a_r^n}$ and $\dfrac{G(n, a_r^n)-G(n, x)}{a_r^n-x}$

therefore

$$\frac{G(n, y)-G(n, x)}{y-x} - f^1(n, a_r^n) = 5\cdot 0(k), \text{ for } n \geqslant k.$$

If on the other hand

$$a_r^n < x < y \leqslant a_{r+1}^n$$

then

$$\frac{G(n, y)-G(n, x)}{y-x} = f^1(n, a_r^n) + \frac{x+y-2a_r^n}{2\Delta_n}\{f^1(n, a_{r+1}^n)-f^1(n, a_r^n)\}$$

so that, again

$$\frac{G(n, y)-G(n, x)}{y-x} - f^1(n, a_r^n) = 5\cdot 0(k), \text{ for } n \geqslant k.$$

Thus for any x, y in (a, b) satisfying $x-y=0(c_{f^1}(k))$ we have

$$\frac{G(n, y)-G(n, x)}{y-x} - f^1(n, x) = 0(k-1), \text{ for } n \geqslant k,$$

proving that $G(n, x)$ is relatively differentiable, with relative derivative $f^1(n, x)$.

Hence $G(n, x)-f(n, x)$ is relatively differentiable with derivative which is zero relative to n, and since $G(n, a)-f(n, a)=0$, therefore, by theorem 3, 2, $G(n, x)$ is equivalent to $f(n, x)$.

We shall show next that, for any x, y in (a, b), $x<y$, there is a (rational) z between x and y such that

$$\frac{G(n, x)-G(n, y)}{x-y}=g(n, z).$$

If $a_r{}^n \leqslant x_1 < x_2 \leqslant a_{r+1}^n (0 \leqslant r < t_n)$ then, clearly,

$$\frac{G(n, x_1)-G(n, x_2)}{x_1-x_2}=g\left(n, \frac{x_1+x_2}{2}\right),$$

and so, if p, q are respectively the least and the greatest integers such that $x \leqslant a_p{}^n \leqslant a_q{}^n \leqslant y$, then

$$\frac{G(n, x)-G(n, y)}{x-y}$$

lies between the least and the greatest of

$$g\left(n, \frac{x+a_p{}^n}{2}\right), \; g\left(n, \frac{a_q{}^n+y}{2}\right) \text{ and } g\left(n, \frac{a_r{}^n+a_{r+1}^n}{2}\right), p \leqslant r < q,$$

and so between the least and the greatest of $g(n, x)$, $g(n, y)$, and $g(n, a_r{}^n)$ for $p \leqslant r \leqslant q$; since $g(n, x)$ is a broken line it follows that there is a rational z between x and y such that

$$\frac{G(n, x)-G(n, y)}{x-y}=g(n, z)$$

(in fact if we rename the points $x, a_r{}^n, y$ as b_r, $p-1 \leqslant r \leqslant q+1$, then $g(n, t)$ is linear in each interval $b_r \leqslant t \leqslant b_{r+1}$; let v be the least integer, not less than $p-1$, such that $g(n, b_v) > \{G(n, x)-G(n, y)\}/(x-y)$ then (unless $v=p-1$), $\{G(n, x)-G(n, y)\}/(x-y)$ lies between $g(n, b_{v-1})$ and $g(n, b_v)$ and is therefore equal to $g(n, z)$ for a rational z between b_{v-1} and b_v, and if $v=p-1$, let μ be the least integer above v for which $g(n, b_\mu) < \{G(n, x)-G(n, y)\}/(x-y)$ then

$$\{G(n, x)-G(n, y)\}/(x-y)$$

lies between $g(n, b_{\mu-1})$ and $g(n, b_\mu)$, etc.

Of course z depends on n. Since $G(n, x)$ and $f(n, x)$ are equivalent, there is a recursive $N(k, x, y)$ such that, for $x \neq y$,

$$\frac{G(n, x)-G(n, y)}{x-y} - \frac{f(n, x)-f(n, y)}{x-y} = 0(k), \text{ for } n \geqslant N(k, x, y);$$

let c_k be the value of z which corresponds to the value $n = N(k, x, y)$ then

$$\frac{f(n, x)-f(n, y)}{x-y} = g(n, c_k) + 0(k),$$

for $n \geqslant N(k, x, y)$. Since $g(n, x)$ is equivalent to $f^1(n, x)$ we conclude that

$$\frac{f(n, x)-f(n, y)}{x-y} = f^1(n, c_k) + 0(k), \text{ for majorant } n.$$

It is important to note that the sequence c_k is not necessarily convergent.

We mention one additional condition on $f(n, x)$ which ensures the convergence of c_k.

If for some constant $\alpha > 0$, and $a \leqslant x < y \leqslant b$,

$$\frac{f^1(n, y)-f^1(n, x)}{y-x} \geqslant \frac{1}{10^{\alpha}}, \text{ for majorant } n,$$

then there is a convergent c_n such that

$$\frac{f(n, b)-f(n, a)}{b-a} = f^1(n, c_n), \text{ relative to } n.$$

For, as we have seen, there is a recursive c_k such that

$$\frac{f(n, b)-f(n, a)}{b-a} = f^1(n, c_k) + 0(k), \text{ for majorant } n.$$

Hence for $N > k$, and majorant n,

$$f^1(n, c_N) - f^1(n, c_k) = 0(k),$$

and therefore

$$c_N - c_k = 0(k-\alpha)$$

proving that c_n is recursively convergent. Since

$$f^1(n, c_n) - f^1(n, c_m) = 0(k)$$

if $m \geqslant c_{f^1}(k) + \alpha$ and n is majorant, and

$$\frac{f(n, b) - f(n, a)}{b - a} = f^1(n, c_m) + 0(m), \text{ for majorant } n,$$

it follows that

$$\frac{f(n, b) - f(n, a)}{b - a} = f^1(n, c_n), \text{ relative to } n.$$

The functions $G(n, x)$, $g(n, x)$ have further important properties which we have not needed to bring out so far; they serve to prove the following

THEOREM 3, 3. 1 *If $f(n, x)$ is relatively differentiable for $a \leqslant x \leqslant b$ then it has an equivalent $G(n, x)$ which is differentiable for $a \leqslant x \leqslant b$ uniformly in x and n.*

With the notation of the previous theorem, let p and q be the least and greatest integers respectively such that if $x \leqslant a_r{}^n \leqslant a_s{}^n \leqslant y$, then $\{g(n, x) - g(n, y)\}/(x - y)$ lies between the least and greatest of

$$\frac{g(n, a^n_{r+1}) - g(n, a_r{}^n)}{a^n_{r+1} - a_r{}^n}$$

for $p - 1 \leqslant r \leqslant q$, and so, since $f^1(n, a^n_{r+1}) - f^1(n, a_r{}^n) = 3 \cdot 0(n)$, therefore

$$\frac{G(n, y) - G(n, x)}{y - x} - g(n, x) = g(n, z) - g(n, x) =$$

$$= 3(z - x) \cdot \frac{0(n)}{\Delta_n} = 3(z - x) \cdot \frac{0(k)}{\Delta_k},$$

if $n \leqslant k$, whence

$$\frac{G(n, y) - G(n, x)}{y - x} - g(n, x) = 0(k), \; n \leqslant k, \; y - x = 0(c_{f^1}(k) + 2).$$

Moreover, from

$$\frac{G(n, y)-G(n, x)}{y-x} = g(n, z),$$

since $g(n, z) - g(n, x) = 0(k-1)$, for $n \geqslant k$ and $y - x = 0(c_{f^1}(k))$, therefore

$$\frac{G(n, y)-G(n, x)}{y-x} - g(n, x) = 0(k)$$

for *all* n, and all x, y in (a, b) satisfying $x-y=0(c_{f^1}(k)+2)$, which proves that $G(n, x)$ is differentiable for $a \leqslant x \leqslant b$, uniformly in x and n.

Since $G(n, x)$ is relatively continuous, it has necessarily a uniformly convergent equivalent, but in fact it is itself uniformly convergent as we shall now show.

Since

$$G(n, a_{r+1}^n)-G(n, a_r^n) = \tfrac{1}{2}\Delta_n \{f^1(n, a_r^n)+f^1(n, a_{r+1}^n)\}$$

therefore,

$$G(n, a_r^n) = f(n, a)+\tfrac{1}{2}\Delta_n \sum_{\mu=0}^{r-1} \{f^1(n, a_\mu^n)+f^1(n, a_{\mu+1}^n)\}.$$

Denoting $r \cdot 10^{c_{f^1}(m)-c_{f^1}(n)}$ by $\sigma(r)$, it follows that, for $m>n$,

$$G(m, a_r^n) - G(n, a_r^n) = G(m, a_{\sigma(r)}^m) - G(n, a_r^n)$$

$$=\{f(m, a)-f(n, a)\}+\tfrac{1}{2}\Delta_m \sum_{\mu=0}^{r-1} \left\{ \sum_{\nu=\sigma(\mu)}^{\sigma(\mu+1)-1} [(f^1(m, a_\nu^m)-f^1(n, a_\mu^n))+ + (f^1(m, a_{\nu+1}^m)-f^1(n, a_{\mu+1}^n))]\right\}$$

$$=0(n)+\tfrac{1}{2}\Delta_m\cdot 8\sigma(r)\cdot 0(n), \text{ for } m \geqslant n,$$

$$=0(n)+4(b-a)\cdot 0(n), \ m \geqslant n.$$

However, for $a_r^n < x \leqslant a_{r+1}^n$,

$$\{G(m, x)-G(n, x)\} - \{G(m, a_r^n)-G(n, a_r^n)\}$$

$$=(x-a_r^n)\left\{\frac{G(m, x)-G(m, a_r^n)}{x-a_r^n} - \frac{G(n, x)-G(n, a_r^n)}{x-a_r^n}\right\}$$

$$=(x-a_r^n)\{g(m, z_1)-g(n, z_2)\}, \text{ where } z_1, z_2 \text{ both lie in } (x, a_r^n)$$

$$=4(x-a_r^n)\cdot 0(n-1), \text{ for } m \geqslant n,$$

and therefore

$$G(m, x) - G(n, x) = 0(k) \text{ for } m \geqslant n \geqslant k + \gamma + 1.$$

THEOREM 3, 3. 2 Rolle's Theorem. *If $f(n, a) = f(n, b)$, relative to n, and if $f(n, x)$ is differentiable in (a, b), relative to n, then we can determine c_k, depending on k alone, such that $a < c_k < b$, and $f^1(n, c_k) = 0(k)$, for majorant n.*

For there is a recursive c_k such that

$$\frac{f(n, b) - f(n, a)}{b - a} = f^1(n, c_k) + 0(k+1), \text{ for majorant } n$$

and by hypothesis

$$\frac{f(n, b) - f(n, a)}{b - a} = 0(k+1), \text{ for majorant } n,$$

whence Rolle's theorem follows.

As with the mean value theorem, we can ensure the convergence of c_k by the condition

$$\frac{f^1(n, x) - f^1(n, y)}{x - y} \geqslant \frac{1}{10^{\alpha}},$$

and when this condition is satisfied, the conclusion in Rolle's theorem takes the form:

$$f^1(n, c_n) = 0, \text{ relative to } n.$$

THEOREM 3, 3. 3 *If $f(n, x), g(n, x)$ are differentiable in (a, b), relative to n, then there is a recursive c_k such that $a < c_k < b$ and*

$$\{f(n, b) - f(n, a)\} g^1(n, c_k) - \{g(n, b) - g(n, a)\} f^1(n, c_k) = 0(k),$$

for majorant n.

We need only apply Rolle's Theorem to the function

$$H(n, x) = \{f(n, b) - f(n, a)\} g(n, x) - \{g(n, b) - g(n, a)\} f(n, x).$$

THEOREM 3, 3. 4 Cauchy's Theorem. *If $f(n, x), g(n, x)$ are differentiable for $a \leqslant x \leqslant b$, relative to n, and if there is a recursive $\alpha(m)$, such that, for sufficiently great values of m,*

$$a + 1/m \leqslant x \leqslant b - 1/m \rightarrow g^1(n, x) \geqslant 10^{-\alpha(m)}$$

then there is a recursive c_k *such that* $a<c_k<b$ *and*

$$\frac{f(n, b)-f(n, a)}{g(n, b)-g(n, a)}=\frac{f^1(n, c_k)}{g^1(n, c_k)}+0(k), \text{ for majorant } n.$$

Applying Theorem 3, 3. 3 to the interval $(a+1/m, b-1/m)$ we determine $c_k{}^m$ between $a+1/m, b-1/m$ such that

$$\{f(n, b-1/m)-f(n, a+1/m)\}\, g^1(n, c_k{}^m)$$
$$-\{g(n, b-1/m)-g(n, a+1/m)\}\, f^1(n, c_k{}^m)=0(k), \text{ for majorant } n.$$

By the mean value Theorem, with $k=\alpha(m)+1$, and $m\geqslant 10^{2-\gamma}$,

$$g(n, b-1/m)-g(n, a+1/m)>10^{\gamma-\alpha(m)-2}$$

whence, writing $\sigma_k{}^m$ for $c^m_{k+2\alpha(m)-\gamma+3}$

$$\frac{f(n, b-1/m)-f(n, a+1/m)}{g(n, b-1/m)-g(n, a+1/m)}=\frac{f^1(n, c_k{}^m)}{g^1(n, c_k{}^m)}+0(k+1), \text{ for majorant } n.$$

It remains to prove that

$$\frac{f(n, b)-f(n, a)}{g(n, b)-g(n, a)}=\frac{f(n, b-1/m)-f(n, a+1/m)}{g(n, b-1/m)-g(n, a+1/m)}+0(k+1),$$

for a suitable choice of m, depending only on k, and for majorant n. This will follow from the relative continuity of $f(n, x)$ and $g(n, x)$, if we can find an absolute positive lower bound for $g(n, b)-g(n, a)$. Let a_1, a_2 be the points of trisection of (a, b) and write $a=a_0$, $b=a_3$; then by the mean value theorem

$$g(n, a_{r+1})-g(n, a_r)=\tfrac{1}{3}(b-a)\{g^1(n, c_{k,r})+0(k)\}, r=0, 1, 2$$

whence, by addition, taking $k=\alpha(m_1)+1$ where m_1 is the least integer which exceeds $3/(b-a)$, we have

$$\begin{aligned} g(n, b)-g(n, a)&=\tfrac{1}{3}(b-a)\{g^1(n, c_{k,1})+g^1(n, c_{k,2})+g^1(n, c_{k,3})+3\cdot 0(k)\}\\ &>\tfrac{1}{3}(b-a)\{g^1(n, c_{k,2})+3\cdot 0(k)\}\\ &>1/m_1 10^{\alpha(m_1)+1} \end{aligned}$$

which furnishes the required lower bound.

3, 3. 1 A function $f(n, x)$ is said to be repeatedly differentiable for $a \leqslant x \leqslant b$, relative to n, if there are recursive functions $f^r(n, x)$, $d_r(k)$ such that $f^0(n, x) = f(n, x)$, and

$$(D)\ a \leqslant x < y \leqslant b \,\&\, x - y = 0(d_r(k)) \rightarrow \{f^r(n, x) - f^r(n, y)\}/(x - y) = \\ = f^{r+1}(x) + 0(k).$$

The function $f^r(n, x)$ is called the rth relative derivative of $f(n, x)$. For each fixed value of r, $f^{r+1}(n, x)$ is the derivative of $f^r(n, x)$ relative to n.

If condition D holds only for $r \leqslant \lambda$, then $f(n, x)$ is said to be $\lambda + 1$ times relatively differentiable.

THEOREM 3, 3. 5 Taylor's Theorem. *If $f(n, x)$ is repeatedly differentiable for $a \leqslant x \leqslant b$, relative to n, and if $g(n, x)$ satisfies the conditions of Cauchy's Theorem, and if $F(n, X, x, r)$ is defined by the recursive equations $F(n, X, x, 0) = f(n, x)$,*

$$F(n, X, r+1) = F(n, X, x, r) + \frac{(X-x)^{r+1}}{(r+1)!} f^{r+1}(n, x)$$

then there is a recursive σ_k in (x, X) such that

$$f(n, X) = F(n, X, x, r) + \frac{g(n, X) - g(n, x)}{g^1(n, \sigma_k)} \cdot \frac{(X - \sigma_k)^r}{r!} f^{r+1}(n, \sigma_k) + 0(k),$$

for majorant n.

We observe first that

$$F^1(n, X, x, r) = \frac{(X-x)^r}{r!} f^{r+1}(n, x);$$

for writing

$$\psi(n, X, x, r) = \frac{(X-x)^r}{r!} f^{r+1}(n, x),$$

we have

$$F^1(n, X, x, 0) - \psi(n, X, x, 0) = 0$$

and since

$$F^1(n, X, x, r+1) = F^1(n, X, x, r) + \\ + \frac{(X-x)^{r+1}}{(r+1)!} f^{r+2}(n, x) - \frac{(X-x)^r}{r!} f^{r+1}(n, x),$$

therefore

$$F^1(n, X, x, r+1) - \psi(n, X, x, r+1) = F^1(n, X, x, r) - \psi(n, X, x, r).$$

Furthermore

$$F(n, X, X, 0) = f(n, X),\ F(n, X, X, r+1) = F(n, X, X, r)$$

and so

$$F(n, X, X, r) = f(n, X).$$

Apply Cauchy's theorem to the functions $F(n, X, x, r)$, $g(n, x)$ in the interval (x, X), then c_k is determined so that

$$\frac{F(n, X, X, r) - F(n, X, x, r)}{g(n, X) - g(n, x)} = \frac{(X - c_k)^r}{r!} \frac{f^{r+1}(n, c_k)}{g^1(n, c_k)} + 0(k),$$

for majorant n.

By relative continuity we can choose k_0 so that

$$|g(n, X) - g(n, x)| < 10^{k_0},$$

and writing $\sigma_k = c_{k+k_0}$, then

$$f(n, X) = F(n, X, x, r) + \frac{g(n, X) - g(n, x)}{g^1(n, \sigma_k)} \frac{(X - \sigma_k)^r}{r!} f^{r+1}(n, \sigma_k) + 0(k),$$

for majorant n.

Taking $g(n, x) = (X - x)^p$ we obtain the Cauchy "remainder"

$$\frac{(X - x)^p (X - \sigma_k)^{r-p+1}}{p(r!)} f^{r+1}(n, \sigma_k),$$

and with $p = r+1$, the Lagrange "remainder"

$$\frac{(X - x)^{r+1}}{(r+1)!} f^{r+1}(n, \sigma_k).$$

3, 4 In Rolle's theorem (and its consequences) the function c_k determined by the theorem satisfies the inequality $a < c_k < b$ and we can determine a numeral m_k such that $a + 1/m_k \leqslant c_k \leqslant b - 1/m_k$; we cannot however determine an m *independent* of k, such that $a + 1/m < c_k < b - 1/m$, for it may happen, for some function $f(n, x)$, as we shall subsequently prove, that however great an m is chosen there may be a value of k for which c_k lies outside the interval

$$(a+1/m, b-1/m)$$

and yet the convergence of c_k to a or b is non-demonstrable.

We shall prove that the function c_k, determined by Rolle's theorem, is *uniformly* contained in $[a, b]$, *i.e.*, that

$$a+1/m \leqslant c_k \leqslant b-1/m,$$

for an m independent of k, provided that the function $f(n, x)$ is *effectively variable* or *effectively constant*, relative to n.

3, 4. 1 A function $f(n, x)$ is *effectively* variable in (a, b), relative to n, if we can determine rationals c_1, c_2 and a numeral α, such that $a \leqslant c_1 < c_2 \leqslant b$ and

$$|f(n, c_1) - f(n, c_2)| \geqslant 1/10^\alpha, \text{ for majorant } n.$$

3, 4. 2 A function $f(n, x)$ is effectively constant in (a, b), relative to n, if $a \leqslant x < y \leqslant b \rightarrow f(n, x) - f(n, y) = 0(k)$, for majorant n.

THEOREM 3, 4 *If $f(n, x)$ is effectively variable in (a, b) relative to n, then we can determine c in (a, b) and a numeral β such that*

$$|f(n, c) - f(n, a)| > 1/10^\alpha, \textit{ for majorant } n.$$

Since

$$|\{f(p, c_1) - f(p, a)\} - \{f(p, c_2) - f(p, a)\}| = |f(p, c_1) - f(p, c_2)| \geqslant 10^{-\alpha}$$

therefore, *either*

$$|f(p, c_1) - f(p, a)| \geqslant 1/10^{\alpha+1} \textit{ or } |f(p, c_2) - f(p, a)| \geqslant 1/10^{\alpha+1},$$

i.e.

$$|f(p, c) - f(p, a)| \geqslant 1/10^{\alpha+1}$$

where c has the value c_1 if the first inequality holds, and the value c_2 otherwise.

But, by convergence,

$$|f(p, c) - f(n, c)| \leqslant 1/10^{\alpha+2}, \; n \text{ majorant}.$$

and therefore

$$|f(n, c) - f(n, a)| \geqslant 1/10^{\alpha+2}, \text{ for majorant } n.$$

The same argument shows also that $|f(n, c) - f(n, b)| \geqslant 1/10^{\alpha+2}$.

THEOREM 3, 4. 1 *If $f(n, x)$ is differentiable in (a, b) relative to n, and if we can determine β, and c in (a, b), such that $|f^1(n, c)| \geqslant 1/10^\beta$, for majorant n, then $f(n, x)$ is effectively variable in (a, b), relative to n.*

We determine c^* in (a, b) so that $|c^* - c| = 1/10^{d(\beta+1)+1}$, then

$$\frac{f(n, c^*) - f(n, c)}{c^* - c} = f^1(n, c) + 0(\beta + 1), \ n \text{ majorant},$$

and so

$$|f(n, c^*) - f(n, c)| \geqslant |c^* - c|/10^{\beta+1} = 1/10^{d(\beta+1)+\beta+2}, \ n \text{ majorant}.$$

THEOREM 3, 4. 2 *If $f(n, x)$ is effectively constant in (a, b), relative to n, then $f(n, x)$ is differentiable, relative to n, with derivative zero.*

For any x, y such that $a \leqslant x < y \leqslant b$,

$$\frac{f(n, y) - f(n, x)}{y - x} = 0(k), \text{ for majorant } n.$$

THEOREM 3, 5 The UNIFORM ROLLE'S THEOREM. *If $f(n, x)$ is differentiable, and effectively variable, or effectively constant, in (a, b), relative to n, and if $f(n, a) = f(n, b)$, relative to n, then we can determine c_k* UNIFORMLY CONTAINED IN THE OPEN INTERVAL $[a, b]$ *such that $f^1(n, c_k) = 0(k)$, for majorant n.*

Suppose, first, that $f(n, x)$ is effectively variable. By Theorem 3, 4 there is a c in (a, b) such that $|f(n, c) - f(n, a)| > 1/10^\beta$ for majorant n; since $f(n, a)$ and $f(n, b)$ are equal, relative to n, it follows that $a < c < b$.

Without loss of generality we may suppose that

$$f(n, c) - f(n, a) > 10^{-\beta} \text{ for majorant } n.$$

By the mean value theorem we can determine c_ϱ^1, c_ϱ^2 in $[a, c]$, $[c, b]$ respectively, such that, for any ϱ,

$$\frac{f(n, c) - f(n, a)}{c - a} = f^1(n, c_\varrho^1) + 0(\varrho),$$

and

$$\frac{f(n, b)-f(n, c)}{b-c} = f^1(n, c_\varrho^{\,2})+0(\varrho), \text{ for majorant } n.$$

Let θ be the greatest integer such that $c-a \leqslant 10^{-\theta}$, $b-c \leqslant 10^{-\theta}$, then, taking

$$\varrho_0 = \max(\beta+1, \beta-\theta+1)$$

we have

$$f^1(n, c^1_{\varrho_0}) > 10^{\theta-\beta-1},\ f^1(n, c^2_{\varrho_0}) < -10^{\theta-\beta-1}.$$

Denoting $c^1_{\varrho_0}$ by $a+1/m_1$, $c^2_{\varrho_0}$ by $b-1/m_2$ then, by the fundamental theorem for relative continuity, Theorem 2, applied to the relatively continuous function $f^1(n, x)$, we can determine c_k in

$$(a+1/m_1, b-1/m_2)$$

such that $f^1(n, c_k)=0(k)$, n majorant.

If $f(n, x)$ is effectively constant then, by Theorem 3, 4. 2, $f^1(n, x)$ is equivalent to zero and in particular $f^1(n, \tfrac{1}{2}(a+b))=0(k)$, n majorant.

Theorem 3, 5. 1 *In Theorem 3. 5 the condition that $f(n, x)$ be effectively variable, or effectively constant, may be replaced by the condition that $f^1(n, x)$ be effectively variable or effectively constant, relative to n.*

For if $f^1(n, x)$ is effectively variable, then, by Theorem 3, 4

$$|f^1(n, c)| > 1/10^\beta,\ n \text{ majorant},$$

and so, by Theorem 3, 4. 1, $f(n, x)$ is effectively variable.

In particular $|f^1(n, c)| > 1/10^\beta$, n majorant, is a sufficient condition for the uniform Rolle's theorem.

On the other hand if $f^1(n, x)$ is effectively constant in (a, b), relative to n, so that $f^1(n, x)-f^1(n, a)=0(k)$ for n majorant, then writing

$$F(n, x)=f(n, x)-xf^1(n, a)$$

we have

$$F^1(n, x)=f^1(n, x)-f^1(n, a)=0(k),\ n \text{ majorant},$$

and therefore, $F(n, b)-F(n, a)=0(k)$, n majorant, whence since $f(n, b)=f(n, a)$, relative to n, we have

$$(b-a)f^1(n, a)=f(n, b)-f(n, a)+0(k)=0(k-1), \text{ for majorant } n.$$

Hence, for $a \leqslant x \leqslant b$,

$$f^1(n, x)=0(k), \text{ for majorant } n.$$

THEOREM 3, 5. 2 THE UNIFORM MEAN-VALUE THEOREM. *If $f(n, x)$ is differentiable in (a, b) relative to n, and if $f^1(n, x)$ is effectively variable or effectively constant, in (a, b), relative to n, then we can determine c_k such that*

$$\frac{f(n, b)-f(n, a)}{b-a}=f^1(n, c_k)+0(k),$$

for majorant n, and c_k is uniformly contained in (a, b)

Write $\varphi(n, x)=f(n, x)-x\left\{\dfrac{f(n, b)-f(n, a)}{b-a}\right\}$.

If $f^1(n, x)$ is effectively variable then there are integers α, V such that

$$|f^1(n, c_1)-f^1(n, c_2)| \geqslant 1/10^\alpha \text{ for } n \geqslant V,$$

and so

$$|\varphi^1(n, c_1)-\varphi^1(n, c_2)| \geqslant 1/10^\alpha \text{ for } n \geqslant V. \tag{i}$$

However

$$\varphi^1(n, x)-\varphi^1(m, x)=0(\alpha+1) \text{ for } n \geqslant m \geqslant \alpha+1;$$

let $\mu=\max(V, \alpha+1)$ then taking μ for n in (i)

$$|\varphi^1(\mu, c)| \geqslant 1/2\cdot 10^\alpha,$$

where c has one of the values c_1, c_2, and therefore

$$|\varphi^1(n, c)| \geqslant 1/4\cdot 10^\alpha \text{ for } n \geqslant \mu.$$

Thus, by Theorem 3, 5. 1, $\varphi(n, x)$ satisfies a sufficient condition for the uniform Rolle's theorem, and Theorem 3, 5. 2 follows.

If $f^1(n, x)$ is effectively constant, then, as in the proof of Theorem 3, 5. 1

$$\{f(n, b)-f(n, a)\}(b-a)=f^1(n, a)+0(k)=f^1(n, x)+0(k-1),$$

for majorant n.

THEOREM 3, 5. 3 THE UNIFORM CAUCHY'S THEOREM. *If* $F(n, x)$, $G(n, x)$ *are differentiable in* (a, b), *relative to* n, *with* $G^1(n, x) \geqslant 1/10^{\alpha(m)}$ *for* $a+1/m \leqslant x \leqslant b-1/m$, $n \geqslant N(m)$, *and if* $F^1(n, x)/G^1(n, x)$ *is effectively variable, or effectively constant in the open interval* $[a, b]$, *then we can determine* c_k *such that*

$$\frac{F(n, b)-F(n, a)}{G(n, b)-G(n, a)} = \frac{F^1(n, c_k)}{G^1(n, c_k)} + 0(k), \text{ for } n \text{ majorant,}$$

and c_k *is uniformly contained in the open interval* $[a, b]$.

Since $F^1(n, x)/G^1(n, x)$ is effectively variable, we can find Q, V and c_1, c_2 in $[a, b]$ such that

$$\left|\frac{F^1(n, c_1)}{G^1(n, c_1)} - \frac{F^1(n, c_2)}{G^1(n, c_2)}\right| \geqslant 1/10^Q \text{ for } n \geqslant V,$$

and so, by uniform convergence, there is a W such that

$$\left|\frac{F^1(n, c)}{G^1(n, c)} - \frac{F(n, b)-F(n, a)}{G(n, b)-G(n, a)}\right| \geqslant 1/10^{Q+1} \text{ for } n \geqslant W,$$

where c has one of the values c_1, c_2.

Since c lies in $[a, b]$, therefore c lies in $(a+1/\mu, b-1/\mu)$ for a certain μ; corresponding to μ we can find R, N such that

$$|G^1(n, x)\{F(n, b)-F(n, a)\}| \geqslant 1/10^R, \text{ for } n \geqslant N$$

and for $a+1/\mu \leqslant x \leqslant b-1/\mu$, and so in particular for $x=c$, and therefore

$$|F^1(n, c)\{G(n, b)-G(n, a)\}-G^1(n, c)\{F(n, b)-F(n, a)\}| \geqslant 1/10^{Q+R+1}$$

for $n \geqslant \max(W, N)$, which proves that the function

$$F(n, x)\{G(n, b)-G(n, a)\}-G(n, x)\{F(n, b)-F(n, a)\}$$

satisfies a sufficient condition for the uniform Rolle's theorem. Accordingly we can find σ_k and ν_k and, independent of k, an m which we take to be not less than μ, such that $a+1/m \leqslant \sigma_k \leqslant b-1/m$ and

$$F^1(n, \sigma_k)\{G(n, b)-G(n, a)\}-G^1(n, \sigma_k)\{F(n, b)-F(n, a)\}=0(k)$$

for $n \geqslant \nu_k$. Hence writing $c_k = \sigma_{k+R}$ we have, for $n \geqslant \nu_{k+R}$,

$$\frac{F(n, b) - F(n, a)}{G(n, b) - G(n, a)} = \frac{F^1(n, c_k)}{G^1(n, c_k)} + 0(k).$$

If on the other hand $F^1(n, x)/G^1(n, x)$ is effectively constant in $[a, b]$, then

$$\frac{F^1(n, x)}{G^1(n, x)} - \frac{F^1(n, c)}{G^1(n, c)} = 0(k), \text{ for majorant } n$$

and for any x, c in $[a, b]$. Since $G^1(n, x)$ is continuous, and so bounded above, relative to n,

$$F^1(n, x)\, G^1(n, c) - F^1(n, c)\, G^1(n, x) = 0(k), \text{ for majorant } n;$$

this result holds for any x in $[a, b]$, and so by continuity for any x in the closed interval (a, b), for majorant n.

Hence

$$\{F(n, b) - F(n, a)\}\, G^1(n, c) - \{G(n, b) - G(n, a)\}\, F^1(n, c) = 0(k),$$

for majorant n, and therefore, taking any c in the interval

$$(a - 1/\mu, b + 1/\mu),$$

it follows that

$$\frac{F(n, b) - F(n, a)}{G(n, b) - G(n, a)} = \frac{F^1(n, c)}{G^1(n, c)} + 0(k)$$

for majorant n, which completes the proof.

THEOREM 3, 5. 4 THE UNIFORM TAYLOR'S THEOREM. *Since Taylor's theorem is an immediate consequence of Cauchy's theorem for the functions*

$$F(n, X, x, r), \; g(n, x)$$

with derivatives (relative to n) $(X-x)^r f^{r+1}(n, x)/r!$, $g^1(n, x)$ respectively, it follows from Theorem 3, 5. 3 *that the uniform Taylor's theorem holds provided that $(X-x)^r f^{r+1}(n, x)/g^1(n, x)$ is effectively variable, or effectively constant. In particular for the Lagrange remainder we have $g(n, x) = (X-x)^{r+1}$, and the condition for the uniform*

Taylor's theorem takes the especially simple form that $f^{r+1}(n, x)$ should be effectively variable, or effectively constant.

3, 5 The uniform Rolle's theorem was established under the additional condition that $f(n, x)$ is effectively variable, relative to n, or effectively constant relative to n, and we pointed out the need for some condition additional to relative differentiability. We shall now show that a proof of the uniform Rolle's theorem without additional qualification, is impossible in recursive analysis. We shall again use the method, which has served in previous chapters, of showing that a proof in recursive analysis of the uniform Rolle's theorem would provides a decision procedure for the class of equations $\varrho(n)=0$, where $\varrho(n)$ is any primitive recursive function which takes only the values 0,1.

Given any recursive $\varrho(n)$, with $\varrho(0)=0$, we define

$$e_0 = 0, \ e_{n+1} = e_n + \prod_{r=0}^{n} (1-\varrho(r))$$

$$d_0 = 1, \ d_{n+1} = 1/e_{n+1}$$

and for $0 \leqslant x \leqslant 1$ and $n \leqslant 3$

$$f(n, x) = \frac{d_n{}^4 x(1-x)}{d_n{}^2+(1-2d_n)x}.$$

We observe first that if the equation

$$e_n = n$$

is provable, then so is the equation

$$\varrho(n) = 0.$$

For from $e_n = n$ follows $e_{n+1} = n+1$ and thence $\prod_{r=0}^{n} (1-\varrho(r))=1$, from which we obtain

$$\varrho(n) = \varrho(n) \prod_{r=0}^{n} (1-\varrho(r)) = 0.$$

We prove readily that $e_n \leqslant n$. For if there is an $r \leqslant n$ such that $\varrho(r)=1$ then $e_n=e_r$, but if $\varrho(r)=0$ for all $r \leqslant n$ then $e_n=n$; hence if there is an $r \leqslant n$ for which $\varrho(r)=1$, and if r_0 is the first, then $e_n=n$ for $n<r_0$ and $e_n=r_0$ for $n \geqslant r_0$.

It follows that if $N>n \geqslant 1$ then

$$0 \leqslant d_n - d_N < 1/n.$$

For if $d_n > 1/n$, then $e_n < n$ and so $e_N = e_n$ and $d_N = d_n$; but if $d_n = 1/n$, then from $d_N \leqslant d_n$ follows

$$0 \leqslant d_n - d_N < 1/n.$$

The magnitude of d_n of course depends upon the existence or non-existence of an n for which $\varrho(n)=1$. In fact we can prove that if there is a recursive $V(k)$ such that

$$n \geqslant V(k) \rightarrow d_n = 0(k), \text{ for all values of } n,$$

then $\varrho(n)=0$ for all n.

For if $e_n \geqslant 10^k$ for $n \geqslant V(k)$, then

$$e_{n+V(n)} \geqslant 10^n > n;$$

since however

$$e_{n+1} = e_n \rightarrow e_{n+p} \leqslant n$$

for any p, we have, taking $V(n)$ for p,

$$e_{n+1} > e_n \text{ for all } n,$$

and therefore $\prod_{r=0}^{n} (1-\varrho(r)) > 0$ for all n, and finally

$$\varrho(n)=0 \text{ for all } n.$$

We show next, that for $n \geqslant 3$ and for $0 \leqslant x \leqslant 1$.

$$0 \leqslant f(n, x) \leqslant d_n^4.$$

Writing, for short, $\lambda = d_n^2/(1-2d_n)$, we have

$$f(n, x) = \lambda d_n^2 \left\{ 2\lambda + 1 - (x+\lambda) - \frac{\lambda(\lambda+1)}{x+\lambda} \right\}$$

whence it follows that (for each n) the rational function $f(n, x)$ attains its maximum value d_n^4 at $x = d_n$.

Furthermore, if $3 \leqslant n < N$ and $0 \leqslant x \leqslant 1$, then

$$f(N, x) \leqslant f(n, x).$$

For

$$\frac{d}{dt}\frac{t^4}{t^2+(1-2t)x} = \frac{2t^3\{t^2+x(2-3t)\}}{\{t^2+(1-2t)x\}^2} > 0,$$

when $0 < t < 2/3$, and d_n is non-increasing.

It follows that, for $3 \leqslant n < N$ and $0 \leqslant x \leqslant 1$,

$$0 \leqslant f(n, x) - f(N, x) < 1/n^4.$$

For if $d_n > 1/n$ then $d_N = d_n$ so that $f(N, x) = f(n, x)$; and if

$$d_n = 1/n \text{ then } 0 \leqslant f(n, x) - f(N, x) < d_n{}^4 = 1/n^4.$$

Thus $f(n, x)$ converges uniformly in x for $0 \leqslant x \leqslant 1$.

A simple calculation shows that, for $0 \leqslant x < X < 1$,

$$\begin{aligned} \left| \frac{f(n, X) - f(n, x)}{X - x} - \lambda d_n{}^2 \left\{ \frac{\lambda(\lambda+1)}{(x+\lambda)^2} - 1 \right\} \right| &= \frac{(X-x)\,\lambda^2(\lambda+1)\,d_n{}^2}{(x+\lambda)^2\,(X+\lambda)} \\ &\leqslant (X-x)(1-d_n)^2 < X - x \end{aligned}$$

which proves that $f(n, x)$ is recursively differentiable in x, uniformly in x and n, and so differentiable relative to n, with derivative

$$f^1(n, x) = \lambda d_n{}^2 \left\{ \frac{\lambda(\lambda+1)}{(x+\lambda)^2} - 1 \right\},$$

provided that $f^1(n, x)$ *converges*. However, for each $n \geqslant 3$, as x increases from 0 to 1, $f^1(n, x)$ steadily decreases from $d_n{}^2$ to $-d_n{}^4/(1-d_n)^2$ and therefore, if $N > n \geqslant 3$, and $0 \leqslant x \leqslant 1$,

$$f^1(N, x) - f^1(n, x) \leqslant d_N{}^2 + d_n{}^4/(1-d_n)^2 < d_N{}^2 + 4d_n{}^4$$

and

$$f^1(N, x) - f^1(n, x) \geqslant -d_N{}^4/(1-d_N)^2 - d_n{}^2 > -4d_N{}^4 - d_n{}^2.$$

If $d_n > 1/n$ then $d_N = d_n$ and $f^1(N, x) = f^1(n, x)$; but if $d_n = 1/n$ then $d_N \leqslant 1/n$, so that in either case

$$-4/n^4 - 1/n^2 < f^1(N, x) - f^1(n, x) < 1/n^2 + 4/n^4$$

proving that $f^1(n, x)$ converges uniformly in x, $0 \leqslant x \leqslant 1$.

Since $f(n, 0) = f(n, 1) = 0$, we have shown that $f(n, x)$ satisfies the conditions of Rolle's theorem.

We come now to the crux of the proof.

Let us assume that (for any $\varrho(n)$) the uniform Rolle's theorem is provable for $f(n, x)$. Then there is a recursive $V(k)$, a recursive c_k and an effectively determined integer $\mathbf{p}$ such that $c_k \geqslant 1/\mathbf{p}$ and

$$n \geqslant V(k) \rightarrow f^1(n, c_k) = 0(k)$$

for all values of n.

Since

$$f^1(n, c_k) = \frac{d_n{}^4(d_n - c_k)\{d_n + (1 - 2d_n)\, c_k\}}{\{d_n{}^2 + (1 - 2d_n)\, c_k\}^2}$$

and

$$d_n + (1 - 2d_n)c_k > d_n{}^2 + (1 - 2d_n)c_k,$$
$$d_n{}^2 + (1 - 2d_n)c_k \leqslant (1 - d_n)^2 < 1,$$

it follows that

$$n \geqslant V(k) \rightarrow d_n{}^4/(d_n - c_k) = 0(k).$$

For the given integer $\mathbf{p}$, either $d_{\mathbf{p}+1} = 1/(\mathbf{p}+1)$ or $d_{\mathbf{p}+1} > 1/(\mathbf{p}+1)$. If $d_{\mathbf{p}+1} = 1/(\mathbf{p}+1)$ then $d_n \leqslant 1/(\mathbf{p}+1)$ for $n \geqslant (\mathbf{p}+1)$ so that

$$|d_n - c_k| > 1/\mathbf{p}(\mathbf{p}+1)$$

and therefore

$$n \geqslant V(4k + \mathbf{p}) \rightarrow d_n = 0(k)$$

for all n, from which it follows, as we have seen, that

$$\varrho(n) = 0$$

for all values of n.

On the other hand, if $d_{\mathbf{p}+1} > 1/(\mathbf{p}+1)$ then there is an r between 0 and $\mathbf{p}+1$ for which $\varrho(r) = 1$. Thus the hypothesis that the uniform Rolle's theorem is provable for $f(n, x)$ implies the existence of a decision procedure for the undecidable class of equations $\varrho(n)=0$.

An important consequence is that *there are recursive functions which are neither relatively variable nor relatively constant*, since we have established the uniform Rolle's theorem for function which have either of these properties.

Of course, the uniform Rolle's theorem may be valid for a function which is neither relatively constant, nor relatively variable.

Let

$$h(n, x)=x(1-x)d_n.$$

Clearly $h(n, x)$ is uniformly convergent in $(0, 1)$ and

$$h(n, 0)=h(n, 1)=0;$$

moreover, for $0 \leqslant x < X \leqslant 1$,

$$\left|\frac{h(n, X)-h(n, x)}{X-x} - (1-2x)\, d_n\right| = (X-x)d_n < (X-x)$$

so that $h(n, x)$ is differentiable in x, uniformly in x and n, with a uniformly convergent derivative $(1-2x)d_n$. Furthermore

$$h^1(n, \tfrac{1}{2})=0 \text{ for all } n,$$

so that the existence of a c_k uniformly contained in $(0, 1)$ is established. However, if $h(n, x)$ were relatively constant there would be a recursive $V(k, x)$ such that, in $(0, 1)$,

$$h(n, x)=0(k), \text{ for } n \geqslant V(k, x);$$

taking $x=1/2$ we have

$$d_n=4\cdot 0(k) \text{ for } n \geqslant V(k, \tfrac{1}{2}),$$

from which it follows that $\varrho(n)=0$ is provable for all n.

And if $h(n, x)$ were relatively variable there is a c in $(0, 1)$ and integers $\mathbf{p}$, $\mathbf{q}$ such that

$$h(n, c) \geqslant 1/\mathbf{p} \text{ for } n \geqslant \mathbf{q}$$

and so, since $h(n, c)$ is the greatest value of $h(n, x)$ in $(0, 1)$ we have

$$d_n \geqslant 4/\mathbf{p} \text{ for } n \geqslant \mathbf{q};$$

let $\boldsymbol{r}$ be the greater of $\boldsymbol{p}$, $\boldsymbol{q}$ then $d_{\boldsymbol{r}} \geqslant 4/\boldsymbol{r}$ so that there is an integer n between 0 and $\boldsymbol{r}$ for which $\varrho(n)=1$.

Thus if $h(n, x)$ were relatively variable or relatively constant for every function $\varrho(n)$ we should again have a decision procedure for the class of equations $\varrho(n)=0$.

Chapter IV

THE RELATIVE INTEGRAL

Ruled functions. Relatively integrable functions. Darboux's Theorem. Continuity of, and derivative of, the relative integral. Substitution in the relative integral.

4 Ruled functions

A recursive function $f(n, x)$ is said to be ruled for $a \leqslant x \leqslant b$ if $f(n, x)$ is uniformly recursively convergent for $a \leqslant x \leqslant b$, and if there are recursive functions $a_r{}^n$, $v_r{}^n$, $b(n)$ and $t(m, n, r)$ such that

$$a_0{}^n = a,\ a^n_{b(n)} = b,\ a^n_{r+1} > a_r{}^n,\ a_r{}^n = a^m_{t(m,n,r)} \text{ for } m > n,$$

and

$$f(n, x) = v_r{}^n, \text{ for } a_r{}^n < x < a^n_{r+1} \text{ and } 0 \leqslant r \leqslant b(n) - 1.$$

A ruled function is absolutely bounded, for if M_0 is the greatest of $|f(0, a_r{}^0)|$, $0 \leqslant r \leqslant b(0)$, and of $|v_r{}^0|$, $0 \leqslant r \leqslant b(0)$, then $|f(0, x)| \leqslant M_0$ for $a \leqslant x \leqslant b$. But, (taking $f(n, x)$ in standard form)

$$|f(n, x) - f(0, x)| < 1$$

and therefore $|f(n, x)| < M_0 + 1$, for all n, and for all x in (a, b). The property of being ruled is not, of course, an invariant of the equivalence relation.

4, 0. 1 In the definition of a ruled function $f(n, x)$ we require $f(n, x)$ to be uniformly convergent. Uniformity is needed to ensure that $f(n, x_n)$ converges when x_n converges, as the following example shows. Let i_n be a nest of intervals whose rational end points tend monotonely to $1/\sqrt{2}$, so that for every rational x in (a, b) x lies outside i_n from some n onwards. Further, let $f(n, x) = n$ in the closed interval i_n, and let $f(n, x) = 0$ outside i_n; for each value of n, $f(n, x)$ is a step function, and for rational x, $f(n, x) = 0$ from some n onwards, so that $f(n, x)$ converges. But if x_n is an end point of i_n, then $f(n, x_n) = n$ so that $f(n, x_n) \to \infty$. Of course $f(n, x)$ is not uniformly convergent, since for any two p, q, with $q > p$, and x in i_q

$$f(q, x) - f(p, x) = q - p \geqslant 1.$$

THEOREM 4, 1 *The sum, difference, product and quotient of two ruled functions are ruled functions.*

For by combining the subdivisions on which the two functions are based, we obtain (for each value of n) a subdivision in each open subinterval of which *both* functions are constant.

THEOREM 4, 1. 1 *A relatively continuous function has a ruled equivalent.*

If $f(n, x)$ is relatively continuous for $a \leqslant x \leqslant b$ and if γ is the least integer such that $b - a \leqslant 10^{\gamma}$, let

$$a_r{}^n = a + (b - a)r/10^{\gamma + c(n)},\ b(n) = 10^{\gamma + c(n)},$$

and

$$\varphi(n, a) = f(n, a),\ \varphi(n, x) = f(n, a_{r+1}^n) \text{ for } a_r{}^n < x \leqslant a_{r+1}^n,$$

then $\varphi(n, x)$ is a ruled equivalent of $f(n, x)$.

THEOREM 4, 1. 2 *If $f(n, x)$ is relatively continuous for $a \leqslant x \leqslant b$ and $g(n, t)$ is a ruled function for $\alpha \leqslant t \leqslant \beta$, and $a \leqslant g(n, t) \leqslant b$ for $\alpha \leqslant t \leqslant \beta$, then $fg(n, t)$ is a ruled function for $\alpha \leqslant t \leqslant \beta$.*

For

$$fg(n, t) = f(n+1, g(c_f(n+1), t)).$$

4, 1 THE RELATIVE INTEGRAL

If $f(n, x)$ is ruled in (a, b) then the sum

$$\sum_{r=0}^{b(n)-1} v_r{}^n (a_{r+1}^n - a_r{}^n)$$

is called a relative integral of $f(n, x)$ from a to b, $(a < b)$, and is denoted by $I_f(n, a, b)$.

If $a = b$ we define $I_f(n, a, b) = 0$, and if $a > b$,

$$I_f(n, a, b) = -I_f(n, b, a).$$

The relative integral $I_f(n, a, b)$ converges recursively. For if $N>n$, and if $t(N, n, r) \leqslant s < t(N, n, r+1)$, then

$$v_s{}^N - v_r{}^n = 0(n)$$

and therefore

$$\begin{aligned} I_f(N, a, b) - I_f(n, a, b) &= \sum_{r=0}^{b(n)-1} \sum_{s=t(N,n,r)}^{t(N,n,r+1)-1} (v_s{}^N - v_r{}^n)(a_{s+1}^N - a_s{}^N) \\ &= (b-a)\cdot 0(n), \end{aligned}$$

which proves that $I_f(n, a, b)$ converges recursively.

Theorem 4, 1. 3 *If $f(n, x)$, $g(n, x)$ are equivalent ruled functions then $I_f(n, a, b)$, $I_g(n, a, b)$ are equivalent.*

By combining the subdivisions on which f and g are constant for a given n, we determine a subdivision $(c_r{}^n)$, say, such that both $f(n, x)$ and $g(n, x)$ are constant in each subinterval $(c_r{}^n, c_{r+1}^n)$; if $\mu_r{}^n$ denotes the mid-point of this subinterval, then

$$I_f(n, a, b) - I_g(n, a, b) = \Sigma\{f(n, \mu_r{}^n) - g(n, \mu_r{}^n)\}\,(c_{r+1}^n - c_r{}^n);$$

but, for $N>n$,

$$f(n, \mu_r{}^n) - f(N, \mu_r{}^n) = 0(n),$$

and

$$g(n, \mu_r{}^n) - g(N, \mu_r{}^n) = 0(n),$$

and

$$f(N, \mu_r{}^n) - g(N, \mu_r{}^n) = 0(n), \text{ for majorant } N,$$

so that

$$I_f(n, a, b) - I_g(n, a, b) = 3(b-a)\cdot 0(n).$$

4, 2 *If a recursive function $f(n, x)$ has a ruled equivalent $F(n, x)$ in (a, b), then $f(n, x)$ is said to be relatively integrable, with a relative integral $I_F(n, a, b)$*; the integral of $f(n, x)$ may also be denoted by $I_f(n, a, b)$. In particular a relatively continuous function is relatively integrable.

Theorem 4, 1. 3 shows that any two relative integrals of $f(n, x)$ are equivalent.

We note that a relatively integrable function is bounded; for if $f(n, x)$ has a ruled equivalent $F(n, x)$, then we know that $|F(n, x)| < M$

for a certain M, and all x and n, and therefore $|f(n, x)| < M+1$ for majorant n.

THEOREM 4, 2 DARBOUX'S THEOREM. *If $f(n, x)$ is relatively integrable in (a, b) and if (x_r), $0 \leqslant r \leqslant N$, is a subdivision of (a, b) and ξ_r any point in (x_r, x_{r+1}), $1 \leqslant r+1 \leqslant N$, then*

$$I_f(k, a, b) - \sum_{r=0}^{N-1} f(n, \xi_r)\,(x_{r+1}-x_r) = 6(b-a)\cdot 0(k),$$

for majorant n, and any subdivision (x_r) of sufficiently small subintervals.

Let M be an absolute bound of a ruled equivalent $F(n, x)$ of $f(n, x)$, in (a, b); by Theorem 4, 1. 3

$$I_f(k, a, b) - I_F(k, a, b) = 3(b-a)\cdot 0(k);$$

since

$$\sum_{r=0}^{N-1} F(n, \xi_r)(x_{r+1}-x_r) - \sum_{r=0}^{N-1} f(n, \xi_r)\,(x_{r+1}-x_r) = 2(b-a)\cdot 0(k)$$

for majorant n, it remains to prove that

$$I_F(k, a, b) - \sum_{r=0}^{N-1} F(k, \xi_r)\,(x_{r+1}-x_r) = (b-a)\cdot 0(k),$$

for suitable subdivisions

Let $(a_r{}^n)$ be the subdivision corresponding to the ruled function $F(n, x)$ (in standard form), so that $F(n, x)$ is constant for

$$a_r{}^n < x < a_{r+1}^n,\ 0 \leqslant r \leqslant b(n),$$

for each value of n, and let $v_r{}^n$ be the value of $F(n, x)$ in the open interval $(a_r{}^n, a_{r+1}^n)$; then

$$I_f(k, a, b) = \sum_{r=0}^{b(k)-1} v_r{}^k(a_{r+1}^k - a_r{}^k).$$

Further let the length of any of the subintervals $x_{r+1}-x_r$, $0 \leqslant r \leqslant N-1$, be less than $a_{r+1}^k - a_r{}^k$ and also less than

$$(b-a)/2M\cdot 10^k\cdot b(k).$$

Finally let c_r, $0 \leqslant r \leqslant p$ be the subdivision formed by combining the two subdivisions (x_r), (a_s^k). In virtue of the limitation on the lengths of the subintervals (x_r, x_{r+1}) at most one point a_s^k falls into any closed interval (x_r, x_{r+1}), $0 \leqslant r \leqslant N-1$.

Let ω_t^k denote the value of $F(k, x)$ in the open interval (c_t, c_{t+1}). Then

$$I_F(k, a, b) = \sum_{t=0}^{p-1} \omega_t^k (c_{t+1} - c_t),$$

and

$$\sum_{r=0}^{N-1} F(k, \xi_r)\,(x_{r+1} - x_r) = \sum_{t=0}^{p-1} F(k, \eta_t)\,(c_{t+1} - c_t)$$

where $\eta_t = \xi_r$ if (c_t, c_{t+1}) is contained in (or coincides with) the interval (x_r, x_{r+1}).

If η_t lies between c_t and c_{t+1} then $f(k, \eta_t) = \omega_t^k$; let $t = t_1, t_2, \ldots, t_q$ be the values of t for which (c_t, c_{t+1}) does not contain η_t. Since an interval (c_t, c_{t+1}) which does not contain η_t can arise only when an a_s^k falls in an interval (x_r, x_{r+1}), it follows that $q \leqslant b(k)$.

Thus

$$\left| \sum_{t=0}^{p-1} F(k, \eta_t)\,(c_{t+1} - c_t) - \sum_{t=0}^{p-1} \omega_t^k (c_{t+1} - c_t) \right|$$

$$= \left| \sum_{r=1}^{q-1} \{F(k, \eta_{t_r}) - \omega_{t_r}^k\}\,(c_{t_r+1} - c_{t_r}) \right|$$

$$\leqslant 2Mb(k) \cdot (b-a)/2M \cdot 10^k \cdot b(k) = (b-a) \cdot 0(k)$$

which completes the proof that

$$I_F(k, a, b) - \sum_{r=0}^{N-1} F(k, \xi_r)\,(x_{r+1} - x_r) = (b-a) \cdot 0(k).$$

THEOREM 4, 2. 1 *If $f(n, x)$ is relatively integrable in (a, b) then $|f(n, x)|$ is relatively integrable in (a, b) and*

$$|I_f(n, a, b)| \leqslant I_{|f|}(n, a, b), \text{ relative to } n.$$

Since $f(n, x)$ is relatively integrable, $f(n, x)$ has a ruled equivalent

$F(n, x)$, and so $|f(n, x)|$ has a ruled equivalent $|F(n, x)|$. Moreover

$$|\Sigma v_r^n(a_{r+1}^n - a_r^n)| \leqslant \Sigma |v_r^n|(a_{r+1}^n - a_r^n)$$

whence the result follows.

THEOREM 4, 2. 2 *If $f(n, x)$ is relatively integrable in (a, b) and if c lies between a and b, then $f(n, x)$ is relatively integrable in (a, c) and (c, b) and*

$$I_f(n, a, c) + I_f(n, c, b) = I_f(n, a, b), \textit{ relative to } n.$$

Let $F(n, x)$ be a ruled equivalent of $f(n, x)$ in (a, b); then of course $F(n, x)$ is a ruled equivalent of $f(n, x)$ in (a, c) and (c, b). Since $I_F(n, a, c) + I_F(n, c, b) = I_F(n, a, b)$ therefore

$$I_f(n, a, c) + I_f(n, c, b) = I_f(n, a, b) + 9(b-a) \cdot 0(n).$$

THEOREM 4, 2. 3 *If $f(n, x)$ and $g(n, x)$ are relatively integrable in (a, b), then $f(n, x) + g(n, x)$ is relatively integrable and*

$$I_f(n, a, b) + I_g(n, a, b) = I_{f+g}(n, a, b), \textit{ relative to } n.$$

Let F, G be ruled equivalents of f, g; then $F+G$ is a ruled equivalent of $f+g$ and

$$I_F(n, a, b) + I_G(n, a, b) = I_{F+G}(n, a, b).$$

The difference between $I_f(n, a, b) + I_g(n, a, b)$ and $I_{f+g}(n, a, b)$ is therefore $9(b-a) \cdot 0(n)$.

THEOREM 4, 3 *If $f(n, x) = 0$ for $a \leqslant x \leqslant b$ and $n \geqslant N$, where N does not depend on x, then $I_f(n, a, b) = 0$ for $n \geqslant N$.*

For $f(n, x)$ is ruled and $\Sigma v_r^n(a_{r+1}^n - a_r^n) = 0$ for $n \geqslant N$.

THEOREM 4, 3. 1 *If $f(n, x)$ is relatively integrable in (a, b) and $f(n, x) \geqslant 0$ in (a, b), then $I_f(n, a, b) \geqslant 0$, relative to n.*

Let $F(n, x)$ be a ruled equivalent of $f(n, x)$, so that

$$F(n, x) > -10^{-k}, \text{ for majorant } n,$$

and therefore

$$F(k, x) > -2/10^k$$

so that

$$I_F(k, a, b) > -2(b-a)/10^k$$

and so

$$I_f(k, a, b) > -5(b-a)/10^k.$$

Similarly if $f(n, x)$ is relatively integrable and $f(n, x) = 0(k)$ in (a, b), for majorant n, then $I_f(n, a, b) = 5(b-a)\cdot 0(k)$, for majorant n.

THEOREM 4, 3. 2 *If $f(n, x)$ is relatively integrable in (a, b) then $I_f(n, a, x)$ is relatively continuous in (a, b).*

If $F(n, x)$ is a ruled equivalent of $f(n, x)$ then $|F(n, x)|$ is bounded by 10^p say, and so for any x, X such that $a \leqslant x < X \leqslant b$, we have

$$|I_F(n, x, X)| < 10^p(X-x) = 0(k), \text{ for } X-x = 0(k+p);$$

but

$$\begin{aligned} I_f(n, a, X) - I_f(n, a, x) &= I_f(n, x, X), \text{ relative to } n, \\ &= I_F(n, x, X), \text{ relative to } n, \\ &= 2\cdot 0(k), \text{ relative to } n, \end{aligned}$$

for $X-x = 0(k+p)$.

THEOREM 4, 4 *If $f(n, x)$ is relatively continuous in (a, b) then $I_f(n, a, x)$ is relatively differentiable in (a, b), with relative derivative $f(n, x)$.*

Let $a \leqslant t < T \leqslant b$ and $\varphi(n, x) = f(n, x) - f(n, t)$, so that if $T-t = 0(c(k))$ and $t \leqslant x \leqslant T$ then

$$\varphi(n, x) = 0(k), \text{ for majorant } n,$$

and so

$$\begin{aligned} &\{I_f(n, a, T) - I_f(n, a, t)\}/(T-t) - f(n, t) \\ &= \{I_\varphi(n, t, T) + 18(b-a)\cdot 0(n)\}/(T-t), \\ &\qquad\qquad \text{by Theorems 4, 2. 2 and 4, 2. 3,} \\ &= 5\cdot 0(k) + \{18(b-a)/(T-t)\}\cdot 0(n) \\ &= 6\cdot 0(k), \text{ for majorant } n. \end{aligned}$$

THEOREM 4, 4. 1 *If $f(n, x)$ has a relative derivative $f^1(n, x)$ in (a, b) then*

$$I_{f^1}(n, a, b) = f(n, b) - f(n, a), \textit{ relative to } n.$$

For

$$I^1_{f^1}(n, a, x) - f^1(n, x) = 0, \text{ relative to } n.$$

4. 3 Substitution in the relative integral

THEOREM 4, 5 *In the interval* (t_0, t_1) *the function* $g(n, t)$ *is relatively differentiable, with a relative derivative* $g^1(n, t)$, *and* $\alpha \leqslant g(n, t) \leqslant \beta$ *for majorant* n. *If* $f(n, x)$ *is relatively continuous for* $\alpha \leqslant x \leqslant \beta$, *and if* $a = g(n, t_0)$, $b = g(n, t_1)$ *relative to* n, *then*

$$I_f(n, a, b) = I_{fg \cdot g^1}(n, t_0, t_1), \text{ relative to } n.$$

From the two conditions (i) $g(n, t_0) = a$, relative to n, (ii) $g(n, t_0) \geqslant \alpha$, it follows that $a \geqslant \alpha$, and similarly that $b \leqslant \beta$, so that $f(n, x)$ is continuous in (a, b), relative to n, and $I_f(n, a, x)$ exists for $a \leqslant x \leqslant b$. Since $fg(n, t)$ and $g^1(n, t)$ are both relatively continuous in (t_0, t_1), therefore the integral $I_{fg \cdot g^1}(n, t_0, t)$ exists for $t_0 \leqslant t \leqslant t_1$.

Writing $F(n, x)$ for $I_f(n, a, x)$, $G(n, t)$ for $Fg(n, t)$ and $H(n, t)$ for $I_{fg \cdot g^1}(n, t_0, t)$, it is readily seen that

$$G^1(n, t) - H^1(n, t) = 0, \text{ relative to } n$$

and therefore

$$G(n, t_1) - G(n, t_0) = H(n, t_1), \text{ relative to } n,$$

whence, since $G(n, t_0) = F(n, a)$ and $G(n, t_1) = F(n, b)$, relative to n, the theorem follows.

4. 4 Relatively inverse functions

THEOREM 4, 6 *In the interval* $0 \leqslant x \leqslant b$, $g(m, x)$ *is differentiable relative to* m, *with a relative derivative* $1/\varphi g(m, x)$, *and*

$$g(m, 0) = \alpha \leqslant g(m, x) \leqslant \beta, \text{ for majorant } m,$$

and in the interval $\alpha \leqslant t \leqslant \beta$, $\varphi(n, t)$ *is continuous relative to* n *and* $\varphi(n, t) \geqslant 10^{-\mu}$, *for majorant* n. *Then* $g(n, x)$ *and* $I_\varphi(n, \alpha, t)$ *are inverse functions, relative to* n.

Denote $I_\varphi(n, \alpha, t)$ by $f(n, t)$; then for $0 \leqslant x \leqslant b$, the relative derivative of $fg(k, x)$ is

$$\varphi g(k, x) \cdot g^1(k, x)$$

which by hypothesis is equal to unity, relative to k. Hence, for $0 \leqslant x \leqslant b$,

4, 4. 1 $$fg(k, x) = x, \text{ relative to } k,$$

and therefore

$$f(n, g(m, x)) = x, \text{ relative to } m, n;$$

by Theorem 3, 2. 5 it follows that, for $\alpha \leqslant t < g(m, x)$, m majorant and $x < b$,

4, 4. 2 $$gf(k, t) = t, \text{ relative to } k.$$

Equations 4, 4. 1, 4, 4. 2 show that f and g are relatively inverse functions.

CHAPTER V

THE ELEMENTARY FUNCTIONS

The relatively exponential, logarithmic and circular functions.

5 The *relatively exponential* function $E(n, x)$ is defined by the recursion

$$E(0, x)=1,\ E(n+1, x)=E(n, x)+x^{n+1}/(n+1)!$$

The function $E(n, x)$ is uniformly recursively convergent in any interval, for if N is any positive integer, and if $|x|<N$ and $n \geqslant 2N$, then

$$\begin{aligned}|E(n+r+1, x)-E(n+r, x)| &\leqslant \{N^n/n!\}\ 2^{-(r+1)}\\ &\leqslant \{N^{2N}/(2N)!\}\ 2^{-(n-2N)}\cdot 2^{-(r+1)}\\ &=\{(2N)^{2N}/(2N)!\}\ 2^{-n}\cdot 2^{-(r+1)}\end{aligned}$$

so that, for $m \geqslant n \geqslant 2N$,

$$|E(m, x)-E(n, x)|<K/2^n,$$

5, 0. 1 where $K=(2N)^{2N}/(2N)!$.

Since, for $|x|<N$, $|X|<N$, we have

5, 0. 2 $$\left|\frac{X^r-x^r}{X-x}\right| \leqslant rN^{r-1}$$

therefore

5, 0. 3 $$\left|\frac{X^{n+1}-x^{n+1}-(X-x)(n+1)x^n}{(X-x)^2}\right| \leqslant$$

$$\leqslant \sum_{r=1}^{n} |x|^{n-r}\left|\frac{X^r-x^r}{X-x}\right| \leqslant N^{n-1}\sum_{r=1}^{n} r$$

$$= n(n+1)\,(2N)^{n-1}/2^n.$$

It follows that

$$\frac{1}{(n+1)!}\left|\frac{X^{n+1}-x^{n+1}}{X-x}-(n+1)\,x^n\right| \leqslant \frac{K}{2^n}\,|X-x|$$

since $(2N)^r/r!$ steadily increases with r up to a maximum for $r=2N$ and then steadily decreases, and therefore if

$$\frac{E(n+1, X)-E(n+1, x)}{X-x} - E(n, x) = F_n$$

we have

$$|F_n - F_{n-1}| < 2^n|X-x|;$$

but $F_0=0$ and therefore

5, 1 $$|F_n| < K|X-x|$$

from which it follows that $E(n, x)$ *is differentiable relative to* n, *with relative derivative* $E(n, x)$, in any interval $-N \leqslant x \leqslant N$.

Accordingly the relative derivative of

$$E(n, -x)\cdot E(n, x+a)$$

is zero relative to n, and therefore

$$E(n, -x)\cdot E(n, x+a)=E(n, 0)\cdot E(n, a)=E(n, a)$$

relative to n; in particular

$$E(n, -x)\cdot E(n, x)=1$$

relative to n, and so

5, 2 $$E(n, x)\cdot E(n, a)=E(n, x+a)$$

relative to n, which is the familiar property of the exponential function.

It follows also from 5, 1 that $E(n, x)$ is continuous in x, uniformly in x and n, in any interval $-N \leqslant x \leqslant N$.

In the same interval

5, 2. 1 $$|E(n, x)| \leqslant \sum_{=0}^{n} N^r/r! = \sum_{r=0}^{n} \frac{(2N)^r}{r!}\cdot\frac{1}{2^r} < 2K$$

where K is given by 5, 0. 1.

5, 3 Denoting the reciprocal function $1/x$ by $R(n, x)$, we define the *relatively logarithmic* function $\log(n, x)$ by the integral

$$\log(n, x)=I_R(n, 1, x), \; x>0.$$

The relative derivative of $\log(n, x)$ is therefore $1/x$.

By Theorem 4, 2. 2

$$I_R(n, 1, a) + I_R(n, a, ab) = I_R(n, 1, ab), \text{ relative to } n,$$

and by Theorem 4, 5, with at for $g(n, t)$ and $R(n, x)$ for $f(n, x)$

$$I_R(n, a, ab) = I_R(n, 1, b), \text{ relative to } n,$$

whence it follows that

5, 3. 1 $\qquad \log(n, a) + \log(n, b) = \log(n, ab)$, relative to n.

5, 4 Since $E(n, x)$ is its own relative derivative, and

$$1/RE(n, x) = E(n, x),$$

therefore in any interval $(0, N)$, $E(m, x)$ has relative derivative $1/RE(m, x)$ and furthermore

$$E(m, 0) = 1 \leqslant E(m, x) \leqslant E(m, N)$$

and for $1 \leqslant t \leqslant E(m, N)$, $R(n, t) \geqslant 1/2K$, so that, by Theorem 4, 6, $E(m, x)$ and $\log(n, x)$ are relatively inverse functions, and for $0 \leqslant x \leqslant N$

5, 4. 1 $\qquad \log(n, E(m, x)) = x$, relative to m, n

and, for $1 \leqslant t \leqslant E(m, N)$ with m majorant, we have

5, 4. 2 $\qquad E(m, \log(n, t)) = t$, relative to n, m.

Since N is arbitrary, and $E(m, N)$ is arbitrarily great with N, therefore 4. 1 holds for all $x \geqslant 0$, and 4.2 holds for all $t \geqslant 1$. However $E(m, -x) = 1/E(m, x)$ relative to m, and $\log(n, 1/t) = -\log(n, t)$ relative to n, so that

$$\log(n, E(m, -x)) = -x, \text{ relative to } m, n$$

for all $x \geqslant 0$, and therefore 4. 1 in fact holds for all x, positive or negative. Writing $1/t$ for t in 4. 2, we see similarly that 4. 2 holds also for $t < 1$, and so for all positive values of t.

5, 5 The *relatively circular functions* $\sin(n, x)$, $\cos(n, x)$ are of course defined by the recursions

$$\sin(0, x) = x,\ \sin(n+1, x) = \sin(n, x) + (-1)^{n+1}x^{2n+3}/(2n+3)!,$$
$$\cos(0, x) = 1,\ \cos(n+1, x) = \cos(n, x) + (-1)^{n+1}x^{2n+2}/(2n+2)!;$$

clearly $\sin(n, 0) = 0$, $\cos(n, 0) = 1$ for all n.

The analysis we made of the function $E(n, x)$ shows that $\sin(n, x)$ and $\cos(n, x)$ are differentiable relative to n, with relative derivatives $\cos(n, x)$ and $-\sin(n, x)$ respectively. One immediate consequence is that

$$\sin^2(n, x) + \cos^2(n, x)$$

has relative derivative zero and therefore

$$\sin^2(n, x) + \cos^2(n, x) = 1, \text{ relative to } n,$$

for all x.

Another consequence is that a relative derivative of the function

$$\cos(n, c-x)\cos(n, x) - \sin(n, c-x)\sin(n, x)$$

is zero and therefore

$$\cos(n, c) = \cos(n, c-x)\cos(n, x) - \sin(n, c-x)\sin(n, x)$$

relative to n, or, writing $x+y$ for c,

5, 5. 1 $\cos(n, x+y) = \cos(n, x)\cos(n, y) - \sin(n, x)\sin(n, y)$

relative to n. Differentiating both sides of this equation as functions of x, we find

5, 5. 2 $\sin(n, x+y) = \sin(n, x)\cos(n, y) + \cos(n, x)\sin(n, y)$

relative to n, which proves the addition formulae.

5, 6 In preparation for later work we find rough bounds for $\sin(n, x)$ and $\cos(n, x)$.

With $0 < |x| \leqslant 4$ and $p \geqslant 1$, we have

$$\{\cos(p+2, x) - \cos(p, x)\}/\{\cos(p+2, x) - \cos(p+1, x)\}$$
$$= 1 - (2p+4)(2p+3)/x^2 < 0,$$

which shows that $\cos(p+2, x)$ lies between $\cos(p, x)$ and $\cos(p+1, x)$, and therefore for $n>p+1\geqslant 2$, and $|x|\leqslant 4$, all $\cos(n, x)$ lie between $\cos(p, x)$ and $\cos(p+1, x)$.

Similarly since

$$\{\sin(p+2, x)-\sin(p, x)\}/\{\sin(p+2, x)-\sin(p+1, x)\} = 1-(2p+4)(2p+5)/x^2<0$$

for the same range of values of x and p, therefore $\sin(n, x)$ lies between $\sin(p, x)$ and $\sin(p+1, x)$, for $n>p+1\geqslant 1$ and $|x|\leqslant 4$. In particular for $0\leqslant x\leqslant 1.6$, $\sin(n, x)$ lies between $\frac{1}{2}x$ and x for $n\geqslant 2$.

In the interval $0\leqslant x\leqslant 3.2$ we have

$$-4.12\leqslant\cos(1, x)\leqslant 1, -0.5\leqslant\cos(2, x)\leqslant 1$$

and therefore all $\cos(n, x)$, $n\geqslant 0$, lie in the interval $(-4.12, 1)$. Similarly, in the same interval, $0\leqslant\sin(0, x)\leqslant 3.2$, and

$$-2.2<\sin(1, x)<1,$$

and so all $\sin(n, x)$, $n\geqslant 0$, lie in the interval $(-2.2, 3.2)$. In the interval $1\leqslant x\leqslant 2$, $\sin(0, x)\geqslant 1$ and $\sin(1, x)\geqslant 2/3$ so that

$$\sin(n, x)\geqslant 2/3 \text{ for all } n.$$

Given any polynomial

$$\sum_{0\leqslant r\leqslant n} a_r x^r = g(x),$$

writing as usual

$$\mathrm{D}g(x) = \sum_{1\leqslant r\leqslant n} r a_r x^{r-1},$$

it follows from Theorems 3, 2.1 and 3, 2.2 that if

$$a\leqslant x\leqslant b \rightarrow m\leqslant \mathrm{D}g(x)\leqslant M$$

then

$$m\leqslant\frac{g(b)-g(a)}{b-a}\leqslant M;$$

(that $g(x)$ is uniformly recursively differentiable follows readily from 5, 0.3).

For fixed p, since $\mathrm{D}\cos(p, x)=\sin(p-1, x)$ and $\mathrm{D}\sin(p, x)= =\cos(p, x)$,

$$\mathrm{D}\{\sin^2(p, x)+\cos^2(p, x)\} = (-1)^p\, 2\cos(p, x)\cdot\frac{x^{2p+1}}{(2p+1)!}$$

and so in the interval $0\leqslant x\leqslant 3.2$,

$$|\mathrm{D}\{\sin^2(p, x)+\cos^2(p, x)\}| < 10\cdot(3.2)^{2p+1}/(2p+1)! < 10^{-2p+20},$$

for

$$\frac{(3.2)^{31}}{31!}=\frac{1}{10^{31}}\cdot\frac{2^{155}}{31!}=\frac{2}{2}\cdot\frac{2}{3}\cdot\frac{4}{4}\cdots\frac{4}{7}\cdot\frac{8}{8}\cdots\frac{8}{15}\cdot\frac{16}{16}\cdots\frac{16}{31}\cdot\frac{2^{57}}{10^{31}}<\frac{(8)^{19}}{10^{31}}<\frac{1}{10^{12}}$$

and so

$$\frac{(3.2)^r}{r!}=\frac{1}{10^r}\cdot\frac{(32)^r}{r!}\leqslant\frac{1}{10^r}\cdot\frac{(32)^{31}}{31!},\quad r\geqslant 31,$$
$$<1/10^{r-19}$$

(in fact $(3.2)^{31}/(31)!$ is much smaller than 10^{-12} but this bound amply suffices for our purpose), whence it follows that

$$|\{\cos^2(p, x)-1\}+\sin^2(p, x)|<(3.2)/10^{2p-20}.$$

At $x=3.1$, we have

$$\sin(5, x) > (3.1)\{1-1.602+1.769-0.176+0.023-0.002\} = = 3.1\times 0.012 > 0$$

and so $\sin(4, 3.1)>0$ and therefore $\sin(n, 3.1)>0$ for all $n\geqslant 4$.

At $x=3.2$ we have

$$\sin(4, x) < (3.2)\{1-1.706+0.874-0.212+0.032\} = = -(3.2)\times 0.012 < 0$$

and therefore $\sin(3, 3.2)<0$, and hence $\sin(n, 3.2)<0$ for all $n\geqslant 3$.

Between $x=3.1$ and $x=3.2$

$$\cos(4, x)<(1+4.37+0.28)-(4.80+1.23)<-1/3$$

whence $\cos(3, x)<-1/3$ and therefore $\cos(n, x)<-1/3$ for all $n\geqslant 3$, which proves that for each n, $\sin(n, x)$ steadily decreases in the interval (3.1, 3.2).

Let $\lambda_0 = 3$, $\lambda_1 = 31$, $\lambda_2 = 314$, $\lambda_3 = 3141$, and for $n \geqslant 4$ let λ_n be the least integer between $31 \cdot 10^{n-1}$ and $32 \cdot 10^{n-1}$ such that

$$\sin(n, (\lambda_n + 1)10^{-n}) < 0$$

and $\sin(n, \lambda_n 10^{-n}) \geqslant 0$; λ_n is primitive recursive. Writing $\pi_n = \lambda_n 10^{-n}$, since $\mathrm{D}\sin(n, x) = \cos(n, x)$ and $|\cos(n, x)| < 5$ for $|x| \leqslant 3.2$, we have

$$0 < \sin(n, \pi_n) - \sin(n, \pi_n + 10^{-n}) < 5 \cdot 10^{-n}$$

so that $0 < \sin(n, \pi_n) < 5 \cdot 10^{-n}$, and therefore

$$-10^{10-2n} < \sin(n+1, \pi_n) < 5 \cdot 10^{-n} + 10^{16-2n}$$

whence $|\sin(N, \pi_n)| < 1/10^{n-1}$ for all N, n such that $N \geqslant n \geqslant 16$.

Since $|\cos(n, x)| \geqslant 1/3$ for $n \geqslant 3$ and $3.1 \leqslant x \leqslant 3.2$, therefore

$$\left| \frac{\sin(N, \pi_N) - \sin(N, \pi_n)}{\pi_N - \pi_n} \right| \geqslant \frac{1}{3}$$

and so

$$|\pi_N - \pi_n| \leqslant 3(10^{1-n} + 10^{1-N}) < 1/10^{n-2}$$

which proves that π_n is primitively recursively convergent. The inequality

$$m \geqslant n \geqslant 16 \rightarrow |\sin(m, \pi_n)| < 1/10^{n-1}$$

shows that $\sin(m, \pi_n) = 0$, relative to n, m and in particular $\sin(n, \pi_n) = 0$ relative to n.

From the relations

$$|\{\cos^2(p, x) - 1\} + \sin^2(p, x)| < 3.2/10^{2p-20},$$

$$N \geqslant n \geqslant 16 \rightarrow |\sin(N, \pi_n)| < 1/10^{n-1}$$

follows

$$N \geqslant n \geqslant 16 \rightarrow |\cos^2(N, \pi_n) - 1| < 1/10^{2n-21}$$

and therefore, since $|1 - \cos(N, \pi_n)| > 1$,

$$|1 + \cos(N, \pi_n)| < 1/10^{2n-21}.$$

5, 6. 1 From the addition formulae it follows that

$$2 \cos^2 (n, \tfrac{1}{2}\pi_n) = 1 + \cos (n, \pi_n), \text{ relative to } n,$$
$$= 0, \text{ relative to } n,$$

so that

$$\cos (n, \tfrac{1}{2}\pi_n) = 0, \text{ relative to } n,$$

and therefore, since $\sin (n, x) \geqslant 2/3$ for $1 \leqslant x \leqslant 2$, and

$$\cos^2 (n, x) + \sin^2 (n, x) = 1, \text{ relative to } n,$$

we have

$$\sin (n, \tfrac{1}{2}\pi_n) = 1, \text{ relative to } n.$$

Furthermore

$$\sin (n, 2\pi_n) = 2 \sin (n, \pi_n) \cos (n, \pi_n), \text{ relative to } n,$$
$$= 0, \text{ relative to } n,$$
$$\cos (n, 2\pi_n) = \cos^2(n, \pi_n) - \sin^2(n, \pi_n), \text{ relative to } n,$$
$$= 1, \text{ relative to } n,$$

whence we find

$$\sin (n, x + 2\pi_n) = \sin (n, x) \cos (n, 2\pi_n) + \cos (n, x) \sin (n, 2\pi_n), \text{ relative to } n,$$
$$= \sin (n, x), \text{ relative to } n,$$

and

$$\cos (n, x + 2\pi_n) = \cos (n, x) \cos (n, 2\pi_n) - \sin (n, x) \sin (n, 2\pi_n), \text{ relative to } n,$$
$$= \cos (n, x), \text{ relative to } n,$$

proving that $\sin (n, x)$ and $\cos (n, x)$ are relatively periodic, with relative period $2\pi_n$.

Since $\cos (n, \tfrac{1}{2} \pi_n - x) = \sin (n, x)$, relative to n, therefore if

$0 < x \leqslant 1.6$ we have $\cos (n, \tfrac{1}{2}\pi_n - x) > \tfrac{1}{3}x$ for majorant n.

5, 7 It is proved, in the appendix, that the recursive real number

(π_n) is primitive recursively transcendental, from which it follows that there is a primitive recursive n_p such that

$$\pi_{p+n_p} \geqslant \pi_p + 1/10^{p+n_p}$$

and we shall use this result in the next section.

5, 7. 1 If $|x| \leqslant \frac{1}{2}\pi_k$ for some k, then

$$\cos(n, x) = \cos(n, \tfrac{1}{2}\pi_n - (\tfrac{1}{2}\pi_n - x)) > \tfrac{1}{6}(\pi_n - \pi_k) > 1/10^l,$$

for a certain l and majorant n.

Hence

$$\tan(n, x) = \sin(n, x)/\cos(n, x)$$

is defined for $|x| \leqslant \frac{1}{2}\pi_k$, bounded above by 10^l for an l which depends primitively recursively upon k, and differentiable relative to n, with relative derivative $1 + \tan^2(n, x)$.

Writing $\varphi(n, t) = 1/(1 + t^2)$, we see that $\varphi(n, t)$ is relatively integrable with a relative integral from 0 to t, denoted by $\arctan(n, t)$. By Theorem 4,6 $\tan(n, t)$ and $\arctan(n, t)$ are relatively inverse functions and for $|x| \leqslant \frac{1}{2}\pi_k$

5, 7. 2 $\qquad \arctan(n, \tan(m, x)) = x$, relative to m, n

and

5, 7. 3 $\qquad \tan(m, \arctan(n, t)) = t$, relative to n, m,

with

$$0 \leqslant t \leqslant \tan(m, x),\ m \text{ majorant}.$$

For $0 < x < 1$, and majorant n,

$$\cos(n, \tfrac{1}{2}\pi_n - x) = \sin(n, x) < 2x,$$
$$\sin(n, \tfrac{1}{2}\pi_n - x) = \cos(n, x) > \tfrac{1}{2}$$

and so

$$\tan(n, \tfrac{1}{2}\pi_n - x) > \tfrac{1}{4}x;$$

in particular,

$$\tan\left(n, \frac{1}{2}\pi_n - \frac{1}{n}\right) \text{ tends to } \infty \text{ as } n \text{ tends to } \infty,$$

so that 7.3 holds for all values of t.

5, 8 The function arcsin (n, x) is defined by the recursion

$$\arcsin(0, x) = x$$
$$\arcsin(n, x) = \arcsin(n-1, x) + \frac{(2n)!}{\{2^n n!\}^2} \frac{x^{2n+1}}{2n+1}, \quad n \geqslant 1$$

The function arcsin (n, x) is recursively convergent for $|x| \leqslant 1$, and in any closed interval I inside the interval $-1<x<1$, the derivative of arcsin (n, x) is uniformly convergent; we denote this derivative by $\varrho(n, x)$ and observe that, in I, $\varrho(n, x)$ is also differentiable with a uniformly convergent derivative $\sigma(n, x)$, say.

Since

$$x \cdot \varrho(n, x) - (1-x^2) \cdot \sigma(n, x) = \frac{(2n+1)!}{\{2^n \cdot n!\}^2} x^{2n+1}$$

therefore, for $|x|<1$,

$$x \cdot \varrho(n, x) = (1-x^2) \cdot \sigma(n, x), \text{ relative to } n.$$

Let $|x| \leqslant \frac{1}{2}\pi_{\mathbf{K}}$ for a certain $\mathbf{K}$; then there is an N such that

$$|x| \leqslant \tfrac{1}{2}\pi_N - 1/2^{N+1}$$

and so

$$|\sin(N, x)| \leqslant \cos(N, 1/2^{N+1}), \text{ relative to } N,$$
$$< 1 - 1/3 \cdot 2^{2N+2}, \text{ for majorant } N,$$

and therefore

$$\sin(m, x) \cdot \varrho(n, \sin(m, x)) = \cos^2(m, x) \cdot \sigma(n, \sin(m, x))$$
relative to m, n;

from which it follows that

$$\varrho(n, \sin(m, x)) \cos(m, x) = 1, \text{ relative to } m, n$$

for $|x| \leqslant \frac{1}{2}\pi_{\mathbf{K}}$, and from this in turn that

$$\arcsin(n, \sin(m, x)) = x, \text{ relative to } m, n.$$

Since $\varrho(n, x) \geqslant 1$ for $x \geqslant 0$, it follows next from Theorem 3, 2. 5 that for $|x|<1$,

$$\sin(m, \arcsin(n, x)) = x, \text{ relative to } n, m$$

which completes the proof that sin (m, x) and arcsin (n, t) are relatively inverse functions.

Chapter VI

TRANSFINITE ORDINALS

In this chapter we construct a part of the theory of transfinite ordinals within recursive arithmetic. The key notion is the expression of an integer in a scale of notation. The sequence of natural numbers, in the scale of 3 for instance,

$$0, 1, 2, 3, 3+1, 3+2, 2.3, 2.3+1, \ldots, 3^2, \ldots, 3^{2.3+2}, \ldots, 3^{3^2}, \ldots, 3^{3^3}, \ldots$$

has an obvious parallel with the sequence of transfinite ordinals

$$0, 1, \ldots, \omega, \omega+1, \ldots, \omega^2, \ldots, \omega^{2\omega+2}, \ldots, \omega^{\omega^2}, \ldots, \omega^{\omega^\omega}, \ldots$$

which we exploit in the following definition.

By means of addition, multiplication and exponentiation any positive integer n can be expressed uniquely in the form

$$c_k s^{a_k} + c_{k-1} s^{a_{k-1}} + \ldots + c_2 s^{a_2} + c_1 s^{a_1} + c_0$$

where $s \geqslant 2$, $0 \leqslant c_0 < s$, $0 < c_1, c_2, \ldots, c_k < s$, $0 < a_1 < a_2 < \ldots < a_k$ and each a_i is itself of this form. If $R_s(n)$ denotes this representation of n with scale symbol s then $R_s(n)$ satisfies the recursive equation

$$R_s(n) = c s^{R_s(a)} + R_s(n - c s^a)$$

where a is the exponent of the greatest power of s which does not exceed n, and cs^a is the greatest multiple of s^a not exceeding n.

We define $T_\omega{}^m(n)$ as the expression obtained by substituting 'ω' for 'm' in the representation of n in the scale of m. For example $T_\omega{}^3(103)$ is the ordinal $\omega^{\omega+1} + 2\omega^2 + \omega + 1$ for $103 = 3^{3+1} + 2.3^2 + 3 + 1$. Every ordinal less than the first ε-number (an ε-number satisfies $\omega^\varepsilon = \varepsilon$) is expressible in the form $T_\omega{}^m(n)$, m being any integer greater than all the integers in the representation of the ordinal; for instance starting with the ordinal

$$\Omega = \omega^{\omega^2} + 7\omega^\omega + 11\omega^{17} + 21$$

we may replace ω by any integer $m > 21$, forming

$$n = m^{m^2} + 7m^m + 11m^{17} + 21$$

and then $\Omega = T_\omega^m(n)$.

The familiar definition of inequality between ordinals makes $\Omega_1 < \Omega_2$ if and only if this inequality holds when, in Ω_1, Ω_2 we replace ω by any sufficiently large integer.

It is not, however, the expression of a number in a scale itself which is of importance in the sequel but the transformation from one number to another by a change of scale. For instance the number 34, expressed in the scale of 3, is $3^3 + 2.3 + 1$, and changing the scale from 3 to 4 (leaving the digits 0, 1, 2 unchanged) yields the number $4^4 + 2.4 + 1 = 265$ so that under a change of scale from 3 to 4, 34 is transformed into 265. Formally we define $T_b^a(n)$, the transform of n under a change of scale from a to b, $b \geqslant a$, as follows.

Let q be the exponent of the greatest power of a contained in n, and let pa^q be the greatest multiple of a^q contained in n, so that p and q are primitive recursive functions of a and n, $a \geqslant 2, n \geqslant 1$. For $b \geqslant a \geqslant 2$, we define:

$$T_b^a(0) = 0,$$

$$T_b^a(n) = pb^{T_b^a(q)} + T_b^a(n - pa^q),\ n \geqslant 1;$$

this definition is a course-of-values recursion so that $T_b^a(n)$ is a primitive recursive function.

To form an ordinal greater than some $T_\omega^m(n)$, we take $T_\omega^M(N)$ such that the integers obtained by substituting sufficiently large integers i for ω, say $T_i^m(n)$ and $T_i^M(N)$, satisfy

$$T_i^m(n) < T_i^M(N)$$

for $i \geqslant m, i \geqslant M$; hence, if $m \geqslant M$, we see that $T_\omega^m(n) < T_\omega^M(N)$ if and only if $T_m^M(N) > T_m^m(n) = n$, i.e. $n < T_m^M(N)$

Thus a decreasing sequence of ordinals takes the form

$$T_\omega^{m_1}(n_1),\ T_\omega^{m_2}(n_2),\ T_\omega^{m_3}(n_3),\ \dots\ ,T_\omega^{m_r}(n_r),\ \dots$$

where, for $r \geqslant 1$, $m_{r+1} \geqslant m_r$, and $n_{r+1} < T_{m_{r+1}}^{m_r}(n_r)$

For a given sequence $m_1, m_2, m_3, \ldots$ we obtain the longest decreasing sequence of ordinals by taking $n_{r+1} = T^{m_r}_{m_{r+1}}(n_r) \dot{-} 1$, since $T_\omega{}^m(n) = 0$ if and only if $n = 0$, and $T^{m_r}_{m_{r+1}}(n) < T^{m_r}_{m_{r+1}}(N)$ if and only if $n < N$.

It was proved by Gentzen that *transfinite* induction over ordinals up to an ε-number is not reducible to ordinary induction, and therefore the decreasing ordinal theorem, which says that every decreasing sequence of ordinals is finite, and which is equivalent to transfinite induction, is not reducible, but transfinite induction over ordinals less than the first ε-number *is* reducible to ordinary induction in a rich enough number system like system Z. Whether or not this reduction can be carried out within some formalisation of recursive arithmetic for all ordinals less than the first ε-number is still an open question but we shall establish this result for ordinals not greater than ω^{ω^ω}.

In virtue of the relationship which have exhibited between transfinite ordinals and the expression of numbers in a scale, the decreasing ordinal theorem, for ordinals less than the first ε-number, is equivalent to the following proposition F^*:
Given any non-decreasing function p_r, with $p_0 \geqslant 2$, an initial n_0, and the function n_r defined by the primitive recursion

$$n_{r+1} = T^{p_r}_{p_{r+1}}(n_r) \dot{-} 1$$

then there is a value of r for which $n_r = 0$.

We start by observing that this proposition is a simple consequence of theorem F:

If $m_{r+1} = T^{p_r}_{p_{r+1}}(m_r \dot{-} 1)$, then there is a value of r for which $m_r = 0$. For if $m_r > 0, r \leqslant s$, and $m_{s+1} = 0$, then taking $n_0 = m_0 \dot{-} 1$, if $n_k = m_k \dot{-} 1$ for some k, we find

$$n_{k+1} = T^{p_k}_{p_{k+1}}(n_k) \dot{-} 1 = T^{p_k}_{p_{k+1}}(m_k \dot{-} 1) \dot{-} 1 = m_{k+1} \dot{-} 1,$$

whence by induction, $n_r = m_r \dot{-} 1$ for all r, and therefore $n_{s+1} = 0$.

We start by proving F for the case $m_0 \leqslant p_0{}^{p_0}$, which is equivalent to proving the ordinal theorem for ordinals $\leqslant \omega^\omega$.

Let $\sigma(n)$ be a non-decreasing function, $\sigma(n) \geqslant 2$; the system of recursive arithmetic in which the following proof will be formalisable will depend upon the nature of $\sigma(n)$ and we shall therefore suppose that $\sigma(n)$ is defined by multiple recursions or at worst by a transfinite recursion of ordinal ω [1])

Let the function $\gamma(a, b, c, n)$ be defined by the primitive recursion

$$\gamma(a, b, c, 0) = (a+1)\{\sigma(c)\}^b$$
$$\gamma(a, b, c, n+1) = T^{\sigma(n+c)}_{\sigma(n+c+1)}\{\gamma(a, b, c, n) \dot{-} 1\},$$

and let $f(x, p, n)$ be defined by the double recursion with parameter substitution

$$f(0, 0, n) = 1 \tag{i}$$
$$f(x+1, p, n) = \varphi(x, p, f(0, p, n), n) \tag{ii}$$
$$f(0, p+1, n) = \varphi(\sigma(n) \dot{-} 1, p, f(0, p, n), n) \tag{iii}$$

where

$$\varphi(a, b, c, d) = c + f(a, b, c+d).$$

The function $f(x, p, n)$ counts the number of terms in the sequence $\gamma(x, p, n, r)$, $r = 0, 1, 2, \ldots$, down to a (not necessarily first) zero. We prove in fact:
for $x, p < \sigma(n)$ and $k \geqslant f(x, p, n)$,

$$\gamma(x, p, n, k) = 0.$$

Denote this proposition by $\mathfrak{F}(x, p, n)$. By equation (i) $\mathfrak{F}(0, 0, n)$ holds, and in virtue of equation (ii)

$$\mathfrak{F}(0, p, n) \,\&\, \mathfrak{F}(x, p, n + f(0, p, n)) \rightarrow \mathfrak{F}(x+1, p, n),$$

for starting from $(x+2)\{\sigma(n)\}^p = (x+1)\{\sigma(n)\}^p + \{\sigma(n)\}^p$ with $x+2 < \sigma(n)$, we reach $(x+1)\{\sigma(n+f(0, p, n))\}^p$ in $f(0, p, n)$ steps and thence we arrive at zero in a further $f(x, p, n+f(0, p, n))$ steps, by hypothesis.

Moreover, starting from

$$\{\sigma(n)\}^{p+1} = \{\sigma(n) \dot{-} 1\}\{\sigma(n)\}^p + \{\sigma(n)\}^p, \; p+1 < \sigma(n),$$

---

[1] See R. Péter, *Rekursive Funktionen*, Budapest 1957

if $\mathfrak{F}(0, p, n)$ and $\mathfrak{F}(\sigma(n) \dot{-} 1, p, n+f(0, p, n))$ hold, then, in virtue of equation (iii), $\mathfrak{F}(0, p+1, n)$ holds. Hence

$$\mathfrak{F}(0, 0, n),$$
$$\mathfrak{F}(0, p, n) \& \mathfrak{F}(x+1, p, n+f(0, p, n)) \rightarrow \mathfrak{F}(x+1, p, n),$$
$$\mathfrak{F}(0, p, n) \& \mathfrak{F}(\sigma(n) \dot{-} 1, p, n+f(0, p, n)) \rightarrow \mathfrak{F}(0, p+1, n),$$

are provable, and from these we may derive $\mathfrak{F}(x, p, n)$ by generalised induction [1]), proving theorem F for $m_0 < p_0^{p_0}$.

The extension to the case $m_0 = p_0^{p_0}$ is simple. Writing σ for $\sigma(n)$, σ_1 for $\sigma(n) \dot{-} 1$, since

$$\sigma^\sigma = \sigma_1 \sigma^{\sigma_1} + \sigma^{\sigma_1}$$

it follows that the sequence $\delta(n, r)$, with

$$\delta(n, 0) = \sigma^\sigma,$$
$$\delta(n, r+1) = T^{\sigma(n+r)}_{\sigma(n+r+1)}\{\delta(n, r) \dot{-} 1\},$$

reaches zero in

$$f(0, \sigma_1, n) + f(\sigma_1, \sigma_1, n+f(0, \sigma_1, n)) = f(0, \sigma, n)$$

steps at most, for $f(0, \sigma_1, n)$ steps lead from

$$\sigma_1 \sigma^{\sigma_1} + \sigma^{\sigma_1} \text{ to } \sigma_1\{\sigma(n+f(0, \sigma_1, n))\}^{\sigma_1},$$

and a further $f(\sigma_1, \sigma_1, n+f(0, \sigma_1, n))$ lead to a zero.

Consider next the sequence $\varepsilon(x, y, z, n, r)$ with

$$\varepsilon(x, y, z, n, 0) = (x+1)\{\sigma(n)\}^{y+z\sigma(n)}$$
$$\varepsilon(x, y, z, n, r+1) = T^{\sigma(n+r)}_{\sigma(n+r+1)}\{\varepsilon(x, y, z, n, r) \dot{-} 1\}.$$

Define the function $f(x, y, z, n)$ by the triple recursion with parameter substitution:

$$f(0, 0, 0, n) = 1 \tag{iv}$$
$$f(x+1, y, z, n) = \varphi(x, y, z, f(0, y, z, n), n) \tag{v}$$
$$f(0, y+1, z, n) = \varphi(\sigma(n) \dot{-} 2, y, z, f(0, y, z, n), n) \tag{vi}$$
$$f(0, 0, z+1, n) = \varphi(\sigma(n) \dot{-} 2, \sigma(n) \dot{-} 1, z, f(0, \sigma(n) \dot{-} 1, z, n), n) \tag{vii}$$

---

1) For the formal details of the derivation see a proof due to P. Bernays in the author's paper [1].

where

$$\varphi(x, y, z, c, n) = c + f(x, y, z, c+n),$$

and let $\mathfrak{F}(x, y, z, n)$ affirm:
if $x, y, z < \sigma(n)$ and $k = f(x, y, z, n)$ then $\varepsilon(x, y, z, n, k) = 0$.
Equation (iv) shows that $\mathfrak{F}(0, 0, 0, n)$ holds.
Since $(x+2)\sigma^p = (x+1)\sigma^p + \sigma^p$, if $f(0, y, z, n)$ steps take one from $\sigma^{y+z\sigma}$ to zero and $f(x, y, z, n)$ steps from $(x+1)\sigma^p$ to zero then by equation (v) $f(x+1, y, z, n)$ steps take one from $(x+2)\sigma^{y+z\sigma}$ to zero, proving

$$\mathfrak{F}(0, y, z, n) \& \mathfrak{F}(x, y, z, n+f(0, y, z, n)) \\ \rightarrow \mathfrak{F}(x+1, y, z, n). \qquad \text{(viii)}$$

Furthermore

$$\sigma^{(y+1)+z\sigma} = \sigma_1 \sigma^{y+z\sigma} + \sigma^{y+z\sigma}$$

and so by equation (vi)

$$\mathfrak{F}(0, y, z, n) \& \mathfrak{F}(\sigma_1 \dot{-} 1, y, z, n+f(0, y, z, n)) \rightarrow \mathfrak{F}(0, y+1, z, n) \quad \text{(ix)}$$

Finally from

$$\sigma^{(z+1)\sigma} = \sigma_1 \sigma^{\sigma_1 + z\sigma} + \sigma^{\sigma_1 + z\sigma}$$

and equation (vii) we deduce

$$\mathfrak{F}(0, \sigma_1, z, n) \& \mathfrak{F}(\sigma_1 \dot{-} 1, \sigma_1, z, n+f(0, \sigma_1, z, n)) \rightarrow \mathfrak{F}(0, 0, z+1, n) \quad \text{(x)}$$

From the proved formulae $\mathfrak{F}(0, 0, 0, n)$ and (viii), (ix), (x) we derive

$$\mathfrak{F}(x, y, z, n)$$

by an application of a generalised induction schema which is also reducible to ordinary induction.

A similar proof[1]) establishes theorem F for a sequence with initial term

$$(x+1)\{\sigma(n)\}^{v_0 + v_1\sigma(n) + v_2\{\sigma(n)\}^2 + \ldots + v_j\{\sigma(n)\}^j}$$

for any definite integer j, and hence for a sequence with initial term σ^{σ^σ}, which is equivalent to proving the decreasing ordinal theorem for ordinals less than or equal to ω^{ω^ω}. The method of proof

[1]) *loc. cit* p. 36.

clearly extends beyond this ordinal but to reach a general ordinal Ω_n, where $\Omega_0=\omega$, $\Omega_{n+1}=\omega^{\Omega_n}$, a quite different approach would appear to be necessary.

If $f(n)$ is an ordinal recursive function of ordinal Ω, so that $f(0)$ is given and

$$f(n)=\varphi(n, f(\lambda(n)))$$

where φ is primitive recursive and $\lambda(n)$ is a predecessor of n in a primitive recursive ordering of the natural numbers of ordinal Ω, then it seems likely that the following result holds (hypothesis T). If R^+ is a formalisation of recursive arithmetic which admits definition by ordinal recursions of any ordinal Ω less than the first ε-number, then the number of terms in a decreasing sequence of ordinals less than or equal to ω^Ω, is given by an ordinal recursive function of ordinal ω^α where $\alpha<\Omega$.

The decreasing ordinal theorem is clearly immediately involved in definition by ordinal recursion for the determination of successive values of $f(n)$ defined by

$$f(n)=\varphi(n, f(\lambda(n)))$$

requires that the sequence $\lambda(n)$, $\lambda(\lambda(n))$, $\lambda(\lambda(\lambda(n)))$, ... should reach 0 in an assignable number of steps, assignable of course by a function which is either primitive recursive, or ordinal recursive for an ordinal less than the ordinal of f itself.

A recent [1]), as yet unpublished, result of Alonzo Church, suggests that hypothesis T may be false for certain ways of representing a decreasing sequence of ordinals in recursive arithmetic, so that the particular interpretation of the decreasing ordinal theorem adopted above may prove to be indispensable for the truth of hypothesis T.

We consider next the extension of the foregoing recursive representation of ordinals beyond the first ε-number. In chapter I we introduced the idea of majorant variables, and relations which held only for majorant variables, i.e. for sufficiently large values

---

1) Communicated to the International Congress of Mathematicians, Edinburgh, 1958.

of the variables. In this section we shall denote majorant variables by ω, ω_r, with $r \geqslant 1$. We recall that a primitive recursive relation $R(n_0, n_1, \ldots, n_k)$ for which there exists a constant c_0 and primitive recursive functions

$$c_{r+1}(n_0, n_1, \ldots, n_r),\ r=0, 1, \ldots, k-1,$$

such that

$$R(n_0, n_1, \ldots, n_r, \ldots, n_k)$$

holds for all $n_0, n_1, \ldots, n_k$ satisfying

$$n_0 \geqslant c_0,\ n_{r+1} \geqslant c_{r+1}(n_0, n_1, \ldots, n_r),\ 0 \leqslant r \leqslant k,$$

is said to hold for majorant $n_0, n_1, \ldots, n_k$.
If $R(n_0, n_1, \ldots, n_k)$ holds for majorant $n_0, n_1, \ldots, n_k$ we write

$$R(\omega, \omega_1, \omega_2, \ldots, \omega_k).$$

For example, given definite numerals a, b, c we have

$$\omega^\omega > a\omega^b + c$$

since $n^n > an^b + c$ if $n > \max(a, b, c)$.
An example with two majorant variables is

$$\omega_1 > \omega^{\omega^\omega}$$

which holds since $N > n^{n^n}$, for all n and all N satisfying

$$N > c_1(n) = n^{n^n}.$$

Another example with two majorant variables is

$$\omega\omega_1{}^\omega + \omega^\omega + 5 > \omega^\omega\omega_1{}^2 + \omega^{\omega^\omega}$$

which holds since

$$nN^n + n^n + 5 > n^nN^2 + n^{n^n},$$

for any $n \geqslant 3$ and $N \geqslant n^{n^n}$.

We start by generalising the concept of a number expressed in a scale of notation with an assigned set of digits.

Let $f(x)$ be a recursive function such that $f(0)=0$, $f(1)=1$ and $f(x+1) \geqslant f(x)+1$ for all x, and let $(k+1)^a$ be the greatest power

of $k+1$ which does not exceed n, and $c(k+1)^a$ the greatest multiple of $(k+1)^a$ which does not exceed n.
Then we define the function $\varphi^f_{k,\sigma}$ by the following recursion:

$$\varphi^f_{k,\sigma}(0)=0$$

$$\varphi^f_{k,\sigma}(n)=f(c)\sigma^{\varphi^f_{k,\sigma}(a)}+\varphi^f_{k,\sigma}(n-c(k+1)^a),\ n\geqslant 1.$$

These equations determine $\varphi^f_{k,\sigma}(n)$ as a function of

$$\sigma,\ f(0),\ f(1),\ \ldots,\ f(k)$$

which we call the representation of n with 'digits' $f(r)$, $0\leqslant r\leqslant k$, and base σ. For example $\varphi^f_{4,\sigma}(10^3)=f(3)\sigma^\sigma+\sigma^{f(4)}$.

When the function f contains more than one argument place the relevant variable for the construction of φ will be placed in the last argument place. If i is the identity function $i(x)=x$, then $\varphi^i_{k,k+1}(n)$ is the familiar representation of n in the scale $k+1$ with digits 0, 1, 2, ..., k.

We note first that

5, 5 $$\varphi^f_{k,\sigma}(r)=f(r),\ \text{if}\ r\leqslant k.$$

For $r=r(k+1)^0$, and so $\varphi^f_{k,\sigma}(r)=f(r)\sigma^{\varphi^f_{k,\sigma}(0)}+\varphi^f_{k,\sigma}(0)=f(r)$.
Similarly

$$\varphi^f_{k,\sigma}(k+1)=\sigma.$$

5, 5. 1 For a fixed $k\geqslant 1$ and a fixed $s\geqslant f(k)+1$, $\varphi^f_{k,s}(n)$ is strictly monotonic increasing with n. Write $\varphi(n)$ for $\varphi^f_{k,s}(n)$; then we have to prove $\varphi(n+1)\geqslant\varphi(n)+1$.

If $n\leqslant k$, this follows from 5,5 since $f(n+1)\geqslant f(n)+1$; suppose that

$$\varphi(n+1)\geqslant\varphi(n)+1$$

holds for $n\leqslant m-1$ where $m\geqslant k+1$. Let c be the exponent of the greatest power of $k+1$ which does not exceed m, b the greatest integer such that $b(k+1)^c\leqslant m$ and let $a=m-b(k+1)^c$, so that $0\leqslant a<(k+1)^c$, $1\leqslant b<k+1$ and $1\leqslant c$. Then, by definition,

$$\varphi(m)=f(b)s^{\varphi(c)}+\varphi(a).$$

We consider in turn the cases $a=0$, $0<a<(k+1)^c-1$, $a+1=(k+1)^c$ and $b+1<k+1$ and finally $a+1=(k+1)^c$, $b=k$.
When $a=0$, $\varphi(a)=0$ and so

$$\varphi(m+1)=f(b)s^{\varphi(c)}+1=\varphi(m)+1.$$

When $0<a<(k+1)^c-1$ then

$$\varphi(m+1)=f(b)s^{\varphi(c)}+\varphi(a+1)\geqslant f(b)s^{\varphi(c)}+\varphi(a)+1=\varphi(m)+1,$$

since $a+1<(k+1)^c\leqslant m-1$ implies $\varphi(a+1)\geqslant\varphi(a)+1$ by hypothesis.
When $a+1=(k+1)^c$ and $b+1<k+1$,

$$\varphi(m+1)=f(b+1)s^{\varphi(c)}\geqslant f(b)s^{\varphi(c)}+s^{\varphi(c)}\geqslant f(b)s^{\varphi(c)}+\varphi(a)+1=\varphi(m)+1$$

for the inequality $a+1=(k+1)^c<m$ implies $1+\varphi(a)\leqslant\varphi(a+1)=s^{\varphi(c)}$.
Finally, when $a+1=(k+1)^c$ and $b=k$ we have

$$\varphi(m+1)=\varphi(\{k+1\}^{c+1})=s^{\varphi(c+1)}\geqslant s^{\varphi(c)}\cdot s\geqslant f(b)s^{\varphi(c)}+$$
$$+\varphi(a)+1=\varphi(m)+1$$

for $m>(k+1)^c\geqslant 1+ck\geqslant 1+c$ so that by hypothesis

$$\varphi(c+1)\geqslant\varphi(c)+1,$$

and, moreover, $s\geqslant f(b)+1$ and $a+1=(k+1)^c<m$ so that

$$s^{\varphi(c)}=\varphi(a+1)\geqslant\varphi(a)+1.$$

Thus the inequality holds for $n\leqslant k$, and if it holds for $n=m-1$ it holds also for $n=m$, and so by induction it holds for all n.

5, 5. 11 In particular, taking $f(x)=i(x)=x$, we have that, for $s>k\geqslant 1$, $\varphi^i_{k,s}(n)$ is strictly increasing with n, and taking

$$f(x)=f^i(x)=\varphi^i_{p,q+1}(x),$$

it follows that if $q\geqslant p\geqslant 1$ and if $s>f^i(k)$ then $\varphi^{f^i}_{k,s}(n)$ is strictly increasing with n.

5, 5. 2 If $f(x)$ and $g(x)$ are strictly monotonic increasing and if $h(x)=f(g(x))$, $p\geqslant g(q)$ and $s>f(p)$ then for $q\geqslant 1$ and all n,

5, 5. 21 $$\varphi^f_{p,s}\{\varphi^g_{q,p+1}(n)\}=\varphi^h_{q,s}(n)$$

For $n=0$, both sides of equation 5.21 are zero; let us suppose that the equation holds for $n=0, 1, \ldots, m-1$. Let a and c be the greatest integers (a chosen first) such that $c\ (q+1)^a \leqslant m$, and write

$$b=m-c(q+1)^a;$$

it follows that $c<q+1$, $b<(q+1)^a$ and $m<(q+1)^{a+1}$. Hence

$$\begin{aligned}\varphi^g_{q,p+1}(m)&=g(c)(p+1)^{\varphi^g_{q,p+1}(a)}+\varphi^g_{q,p+1}(b)\\ &<g(c)(p+1)^{\varphi^g_{q,p+1}(a)}+(p+1)^{\varphi^g_{q,p+1}(a)+1},\end{aligned}$$

by 5.1, since $b<(q+1)^a$,

$$=\{g(c)+1\}(p+1)^{\varphi^g_{q,p+1}(\varphi)}<(p+1)^{\varphi^g_{q,p+1}(a)+1},$$

since $g(c)\leqslant g(q)<p+1$.

This shows that $\varphi^g_{q,p+1}(a)$ is the exponent of the greatest power of $p+1$, and $g(c)$ the multiplier of the greatest multiple of this power which does not exceed $\varphi^g_{q,p+1}(m)$. Therefore

$$\begin{aligned}\varphi^f_{p,s}\{\varphi^g_{q,p+1}(m)\}&=h(c)s^{\varphi^f_{p,s}\{\varphi^g_{q,p+1}(a)\}}+\varphi^f_{p,s}\{\varphi^g_{q,p+1}(b)\}\\ &=h(c)s^{\varphi^h_{q,s}(a)}+\varphi^h_{q,s}(b), \text{ by the induction hypothesis,}\\ &=\varphi^h_{q,s}(m), \text{ since } m=c(q+1)^a+b,\end{aligned}$$

which completes the proof of 5.21 by induction.

5, 5. 3 We define next a function $X_n(k)$ which depends upon two functions $p(n)$, $\pi(n)$; $p(n)$ and $\pi(n)$ are arbitrary recursive functions with $p(n)\geqslant 1$.

We define

$$X_0(k)=k$$

$$X_{n+1}(k)=\varphi^{X_n}_{p(n),\,X_n(p(n))+\pi(n)+1}(k).$$

Writing $p'(n)=X_n(p(n))+\pi(n)$, we have

$$X_{n+1}(k)=\varphi^{X_n}_{p(n),\,p'(n)+1}(k), \text{ for all } n.$$

When it is necessary to show the dependance of X upon p, p' explicitly we shall write $X^{p,p'}_n(k)$ for $X_n(k)$.

To justify the definition of $X_n(k)$ we must show that for all $n \geqslant 0$, $X_n(k)$ is monotonic increasing with k.

This is certainly true for $n=0$, since $X_0(k)=k$, and if it is true for $n=m$, then it holds for $n=m+1$, by 5.1.

It follows that, for all n and k, $X_n(k) \geqslant k$.

5, 5. 31 We consider next the relationship between $X^{p,p'}$, $X^{p,q}$ and $X^{q,p'}$. We shall prove that, if $p(r) \geqslant 1$, $q(r) \geqslant 1$,

$$p'(r) \geqslant \max \{X_r{}^{p,p'}(p(r)), X_r{}^{q,p'}(q(r))\},$$

and

$$q(r) \geqslant X_r{}^{p,q}(p(r))$$

for all r, then

5, 5. 4 $$X_n{}^{p,p'}(k) = X_n{}^{q,p'}(X_n{}^{p,q}(k))$$

for all n, k.

Since $X_0{}^{s,t}(k)=k$ for any s, t therefore 5.4 holds for $n=0$; if it holds for $n=m$ then

$$\begin{aligned} X^{q,p'}_{m+1}(X^{p,q}_{m+1}(k)) &= \varphi^{X^{q,p'}_m}_{q(m),p'(m)+1}(X^{p,q}_{m+1}(k)) \\ &= \varphi^{X^{q,p'}_m}_{q(m),p'(m)+1}(\varphi^{X^{p,q}_m}_{p(m),q(m)+1}(k)) \\ &= \varphi^{X^{p,p'}_m}_{p(m),p'(m)+1}(k), \\ &= X^{p,p'}_{m+1}(k), \text{ by 5.21,} \end{aligned}$$

since, by the induction hypothesis,

$$X_m{}^{p,p'}(x) = X_m{}^{q,p'}(X_m{}^{p,q}(x)).$$

Thus 5.4 holds for $n=m+1$, and so for any n.

5, 5. 5 The relationship which holds between $X_n{}^{p,r}(p(n))$ and $X_n{}^{q,r}(q(n))$ persists for all sufficiently large r. More precisely if $p(k) \geqslant 1$, $q(k) \geqslant 1$,

$$r(k) \geqslant \max \{X_k{}^{p,r}(p(k)), X_k{}^{q,r}(q(k))\}$$

and

$$s(k) \geqslant \max \{X_k{}^{p,s}(p(k)), X_k{}^{q,s}(q(k))\}$$

then, for $n > k \geqslant 0$,

$$X_n{}^{p,r}(p(n)) \gtreqless X_n{}^{q,r}(q(n))$$

according as

$$X_n^{p,s}(p(n)) \gtreqless X_n^{q,s}(q(n)).$$

Define

$$t(n),\ X_n^{p,t},\ X_n^{q,t},\ X_n^{r,t}$$

and $X_n^{s,t}$ simultaneously by the recursions

$$X_0^{p,t}(k)=X_0^{q,t}(k)=X_0^{r,t}(k)=X_0^{s,t}(k)=k,$$

$$t(n)= \max \{X_n^{p,t}(p(n)),\ X_n^{q,t}(q(n)),\ X_n^{r,t}(r(n)),\ X_n^{s,t}(s(n))\}$$

and

$$X_{n+1}^{p,t}=\varphi^{X_n^{p,t}}_{p(n),t(n)+1},\quad X_{n+1}^{q,t}=\varphi^{X_n^{q,t}}_{q(n),t(n)+1},$$

$$X_{n+1}^{r,t}=\varphi^{X_n^{r,t}}_{r(n),t(n)+1},\quad X_{n+1}^{s,t}=\varphi^{X_n^{s,t}}_{s(n),t(n)+1}.$$

Then, by 5.4,

$$X_n^{r,t}(X_n^{p,r}(p(n))=X_n^{p,t}(p(n)))$$

and

$$X_n^{r,t}(X_n^{q,r}(q(n))=X_n^{q,t}(q(n)))$$

and therefore by 5.3,

$$X_n^{p,t}(p(n)) \gtreqless X_n^{q,t}(q(n))$$

according as

$$X_n^{p,r}(p(n)) \gtreqless X_n^{q,r}(q(n)).$$

Similarly

$$X_n^{p,t}(p(n)) \gtreqless X_n^{q,t}(q(n))$$

according as

$$X_n^{p,s}(p(n)) \gtreqless X_n^{q,s}(q(n))$$

from which 5.5 follows.

5, 5. 6 If $p(n) \geqslant k > 0$, and for $0 \leqslant r \leqslant n$, $p(r) \geqslant 1$, $p'(r) \geqslant X_r^{p,p'}(p(r))$, then

$$X_n^{p,p'}(k)=X_{n+1}^{p,p'}(k).$$

For $X_{n+1}^{p,p'}(k)=\varphi^{X_n^{p,p'}}_{p(n),p'(n)+1}(k)=X_n^{p,p'}(k)$, since $k \leqslant (p(n))$.

Hence if $N > n$ and, for $n \leqslant m < N$, $p(m) \leqslant k$, then

$$X_n^{p,p'}(k)=X_N^{p,p'}(k).$$

5, 6 We define

$$\Omega_0^p(k)=k,$$

$$\Omega_{n+1}^p(k)=\varphi_{p(n),\omega_n}^{\Omega_n^p}(k), \text{ with } \omega_0=\omega,$$

where ω, ω_r, $r\geqslant 1$, are majorant variables.
For given k, n and $p\ (r)$ these equations determine $\Omega_n{}^p(k)$ as a function of $\omega, \omega_1, \ldots, \omega_{n-1}$ which we call a *transfinite ordinal of type n*.

A transfinite ordinal of type n is equal to an ordinal of type $n+1$ for, if $p(n)\geqslant k$, then

5, 6. 1
$$\Omega_{n+1}^p(k)=\varphi_{p(n),\omega_n}^{\Omega_n^p}(k)=\Omega_n{}^p(k).$$

It follows that if $p(m)\geqslant k$ for $n\leqslant m\leqslant N-1$ then

5, 6. 2
$$\Omega_n{}^p(k)=\Omega_N{}^p(k).$$

In formulating the definition of $\Omega_n{}^p(k)$ we have anticipated a proof that $\Omega_n{}^p(k)$ is strictly increasing with k; this follows from $\Omega_0{}^p(k)=k$ and theorem 5. 1.

5, 6. 3 It is readily seen that there are ordinals of type $n+1$ which are not of lower type. For if $k>p(n)\geqslant m$, then

$$\Omega_{n+1}^p(k)>\Omega_{n+1}^p(m)=\Omega_n{}^p(m).$$

The relationship between ordinals of different types is given by Theorem

5, 6. 4 if $p(r)\geqslant 1$, $q(r)\geqslant X_r{}^{p,q}(p(r))$ for $0\leqslant r\leqslant n-1$, then

$$\Omega_n{}^p(k)=\Omega_n{}^q(X_n{}^{p,q}(k)).$$

We proceed again by induction. For $k=0$, the equality follows immediately from the definitions; if it holds for $n=m$ then

$$\Omega_{m+1}^q(X_{m+1}^{p,q}(k))=\varphi_{q(m),\omega_m}^{\Omega_m^q}(\varphi_{p(m),q(m)+1}^{X_m^{p,q}}(k))$$
$$=\varphi_{p(m),\omega_m}^{\Omega_m^p}(k),$$

by hypothesis, and therefore

$$\Omega^q_{m+1}(X^{p,q}_{m+1}(k)) = \Omega^p_{m+1}(k),$$

proving the equality for $n=m+1$, and so for all n.

5, 6. 5 If $p(r)\geqslant 1$, $q(r)\geqslant 1$, $0\leqslant r\leqslant n$ then

$$\Omega_n{}^p(p(n)) \gtreqless \Omega_n{}^q(q(n))$$

according as

5, 6. 51 $$X_n{}^{p,r}(p(n)) \gtreqless X_n{}^{q,r}(q(n))$$

5, 6. 52 where $r(k)\geqslant \max\{X_k{}^{p,r}(p(k)), X_k{}^{q,r}(q(k))\}$ for $n>k\geqslant 0$.

For, by 5, 6. 4,

$$\Omega_n{}^p(p(n)) = \Omega_n{}^r(X_n{}^{p,r}(p(n)))$$

and

$$\Omega_n{}^q(q(n)) = \Omega_n{}^r(X_n{}^{q,r}(q(n))).$$

whence 6.5 follows from 6.3.

By 5, 5. 5 the relation 6.51 is independent of the choice of the numbers $r\ (k)$ when condition 6. 52 is satisfied.

5, 7 The result contained in 6.5 enables us to extend the *statement* of the decreasing ordinal theorem to ordinals of any type. In terms of a sequence of functions $p_n(k)$ we define

$$\Lambda^i{}_s(n) = X_i{}^{p_n,s}(p_n(i)).$$

The sequence of ordinals $\Omega_i{}^{p_n}(p_n(i))$, $n=0, 1, 2, \ldots$ is steadily decreasing if

$$\Lambda^i{}_{r_n}(n+1) < \Lambda^i{}_{r_n}(n)$$

where

$$r_n(k) \geqslant \max\{\Lambda_{r_n}{}^k(n), \Lambda_{r_n}{}^k(n+1)\},\ 0\leqslant k<i.$$

Hence the theorem that a decreasing sequence of ordinals is necessarily finite may be expressed in the form:

If for a given integer i, $p_n(i)\geqslant 0$, and for $0\leqslant k<i$, $p_n(k)\geqslant 1$ and if

$$r_n(k) = \max\{\Lambda_{r_n}{}^k(n), \Lambda_{r_n}{}^k(n+1)\}$$

and

$$\Lambda_{r_n}{}^i(n+1) < \Lambda_{r_n}{}^i(n)$$

when $p_n(i) > 0$, then there exists a value of n for which $p_n(i) = 0$
5, 8 By introducing additional operations, transcending exponentiation, we may define ordinals of greater "complexity" than $\Omega_n{}^p(k)$. For instance if we define *tetration* ${}^b a$ so that ${}^0 a = 1$,

$$^{(n+1)}a = a^{({}^n a)},$$

and so ${}^1 a = a$, ${}^2 a = a^a$, ${}^3 a = a^{a^a}$, and so on,) then we may generalise representation in a scale to four operations. Take, a, b, c in turn to be the greatest integers such that (for $k \geqslant 1$)

$$^{a}(k+1) \leqslant n,$$
$$\{^{a}(k+1)\}^{b} \leqslant n,$$
$$c\{^{a}(k+1)\}^{b} \leqslant n,$$

and define

$$\varphi^f_{4,k,s}(n) = f(n),\ 0 \leqslant n \leqslant k,$$
$$\varphi^f_{4,k,s}(n) = \varphi^f_{4,k,s}(c)\{^{\varphi^f_{4,k,s}(a)}s\}^{\varphi^f_{4,k,s}(b)} + \varphi^f_{4,k,s}(n - c\{^{a}(k+1)\}^{b}),\ n \geqslant k+1.$$

The function $\varphi^f_{4,k,s}(n)$ determines the representation of n with digits $f(0), f(1), \ldots, f(k)$, base s, using four operations, addition, multiplication, exponentiation and tetration.
Taking $f(x) = x$, we have for instance

$$\varphi^f_{4,k,s}(10^3) = (\omega+1)\{^{\omega+1}\omega\}^{\omega} + (\omega+1)\{^{\omega}\omega\}\{^{\omega+1}\omega\} + \omega\{^{\omega+1}\omega\} + \omega\{^{\omega}\omega\},$$

(since $1000 = (1+2)\{2^{2^2}\}^2 + (1+2)2^2 \cdot 2^{2^2} + 2.2^{2^2} + 2.2^2$).
Using p operations instead of four we define,

$$G(k+1, a, n+1) = G(k, a, G(k+1, a, n))$$

with $G(3, a, 0) = 1$, $G(2, a, 0) = 0$, $G(1, a, 0) = a$, $G(0, a, n) = n+1$,

$$g^h_{p,k}(1) = G(p, k+1, h(1))$$
$$g^h_{p,k}(m+1) = G(p-m, g^h_{p,k}(m), h(m+1)),\ 1 \leqslant m \leqslant p-1;$$

next, for $n \geqslant k+1$, we determine $a(r)$ so that for $1 \leqslant r \leqslant p$, $a(r)$ is the greatest integer such that, for $1 \leqslant s \leqslant r$,

$$g^a_{p,k}(s) \leqslant n,$$

(and therefore $a(r) < n$, $1 \leqslant r \leqslant p$).

Finally we define

$$Y^f_{p,k,s}(1)=G(p,\ s,\ \varphi^f_{p,k,s}(a(1))),$$

$$Y^f_{p,k,s}(m+1)=G(p-m,\ Y^f_{p,k,s}(m),\ \varphi^f_{p,k,s}(a(m+1))),\ 1\leqslant m\leqslant p-1,$$

and

$$\varphi^f_{p,k,s}(n)=f(n),\ n\leqslant k$$

$$\varphi^f_{p,k,s}(n)=Y^f_{p,k,s}(p),\ n\geqslant k+1.$$

APPENDIX

RECURSIVE IRRATIONALITY AND TRANSCENDENCE

Let $P_n(x)$ be the nth polynomial in an enumeration of all polynomials of a single variable x, with integral coefficients; let

$$||z|| = ||x+iy|| = |x|+|y|$$

and let s_n be a convergent sequence of rational real or complex numbers. Then classically $\lim s_n$ is transcendental if

1. $$(\forall r)(\exists k)(\exists N)(\forall n)\{n \geqslant N \to ||P_r(s_n)|| > 2^{-k}\}$$

The convergence of s_n is expressed by the condition

2. $$(\forall k)(\exists \nu)(\forall n)\{n \geqslant \nu \to ||s_n - s_\nu|| < 2^{-k}\}.$$

Let $\nu(k)$ be the least value of ν, given by (2), such that

$$n \geqslant \nu(k) \to ||s_n - s_{\nu(k)}|| < 2^{-(k+2)}$$

and let k_r, N_r be the least values of k, N given by (1) such that

3. $$n \geqslant N_r \to ||P_r(s_n)|| > 2^{-k_r}.$$

If s_n is quasi-recursive, and recursively convergent, so that $\nu(k)$ is recursive, and if further the functions N_r, k_r in (3) are both quasi-recursive then the recursive real (complex) number (s_n) is said to be *quasi recursively transcendental.*

If s_n, $\nu(k)$, N_r and k_r are all *primitive recursive* then the primitive recursive real number (s_n) is said to be *primitive recursively transcendental.*

In particular taking $P_r(x)$ to be a linear function of x we obtain the corresponding definitions of recursive irrationality.

For a quasi-recursive, recursively convergent sequence s_n, the number (s_n) is quasi-recursively transcendental if and only if it is classically transcendental.

For if $M = \max_{0 \leqslant r \leqslant \nu(1)} ||s_r|| + 1$ and if $P_r^*(x)$ is the sum of the positive values of the terms of the derived polynomial $P_r'(x)$ then

$$||P_r(s_m) - P_r(s_n)|| < ||s_m - s_n|| \cdot P_r^*(M),$$

and calling c_r the exponent of the least power of 2 which exceeds $P_r^*(M)$ we have

4. $$m, n \geqslant \nu(k + c_r) \rightarrow ||P_r(s_m) - P_r(s_n)|| < 2^{-k}.$$

From (3) and (4) taking k_r for k and $N_r + \nu(k_r + c_r + 2)$ for n, we find, writing $\nu_r(k)$ for $\nu(k_r + c_r + 2)$,

$$||P_r(s_{\nu r(k_r)})|| > 1/2^{k_r+1}$$

whence

5. $$(\forall r)(\exists k)\{||P_r(s_{\nu r(k)})|| > 1/2^{k+1}\}.$$

If λ_r is the least value of k satisfying (5) then since $\nu(k)$ is quasi-recursive, so is λ_r, and

$$||P_r(s_{\nu r(\lambda_r)})|| > 1/2^{\lambda_r+1}.$$

Using (4) again, with $k = \lambda_r + 2$, we have

$$n \geqslant \nu_r(\lambda_r) \rightarrow ||P_r(s_n)|| > 1/2^{\lambda_r+2}$$

which proves that (s_n) is quasi-recursively transcendental. This theorem does not of course remain true if we replace quasi by primitive recursion.

For any rational $x \neq 0$, $e^x = \Sigma(x)^n/n!$ is primitively recursively convergent [1] and so is π.[2]
Both e and π are in fact primitive recursively transcendental as we shall now show.

THE PRIMITIVE RECURSIVE TRANSCENDENCE OF e

Let $\varphi(x) = \dfrac{x^{p-1}}{(p-1)!} \prod_{r=1}^{n} (r - x)^p = \sum_{s=p-1}^{q} c_s x^s$, where $q = pn + p - 1$,

---

1) *Vide* R. L. Goodstein [3].
2) *Vide* R. L. Goodstein [12].

and

$$\psi(x) = \sum_{s=0}^{n} a_s x^s, \; A = \max_{1 \leqslant s \leqslant n} |a_s|;$$

further let

$$T\varphi(x) = \sum_{s=0}^{q} \varphi^s(x)$$

where

$$\varphi^s(x) = \sum_{r=0}^{q-s} \frac{(r+s)!}{r!} c_{r+s}\, x^r, \; 0 \leqslant s \leqslant q.$$

Then, by elementary algebra, we may readily show that

$$|\varphi^{p-1}(0)| = (n!)^p,$$

and that for $r \neq p-1$, $\varphi^r(0) = 0 \pmod{p}$, and for $1 \leqslant m \leqslant n$, and any r, $\varphi^r(m) = 0 \pmod{p}$, so that if p is prime and greater than n, then $T\varphi(0)$ is not divisible by p, but $T\varphi(m) = 0 \pmod{p}$, $1 \leqslant m \leqslant n$. Now $T\varphi(0) = \sum_{r=p-1}^{q} (r!)c_r$ and so, writing $E_N(m)$ for $\sum_{r=0}^{N} m^r/r!$,

$$T\varphi(0)E_N(m) = \sum_{p-1}^{q} c_r\{Tm^r + W_r m^r\}, \; N > q,$$

where

$$Tm^r = m^r + rm^{r-1} + r(r-1)m^{r-2} + \ldots + r!$$

and

$$W_r = \frac{m}{r+1} + \frac{m^2}{(r+1)(r+2)} + \ldots + \frac{(r!)m^{N-r}}{N!} < 3^m,$$

$\left(\text{for } 1 + \frac{m}{1!} + \frac{m^2}{2!} + \ldots + \frac{m^n}{n!} < \left\{1 + \frac{1}{1!} + \frac{1}{2!} + \ldots + \frac{1}{n!}\right\}^m\right.$, as may readily be shown by induction over $\left.m\right)$.

Hence

$$T\varphi(0) \cdot E_N(m) = T\varphi(m) + 3^m \sum_{r=p-1}^{q} c_r \Theta_r m^r,$$

where

$$\Theta_r = W_r 3^{-m}, \text{ so that } 0 < \Theta_r < 1.$$

But

$$\left|\sum_{r=p-1}^{q} c_r \Theta_r m^r\right| \leqslant \sum_{r=p-1}^{q} |c_r| m^r = \frac{m^{p-1}}{(p-1)!} \prod_{r=1}^{n} (r+m)^p$$

$$< \frac{M^p}{(p-1)!}, \ 1 \leqslant m \leqslant n,$$

where $M = \prod_{r=0}^{n} (r+n)$.

Now if $p > 2M$

$$\frac{M^p}{(p-1)!} < \frac{(2M)^{2M}}{(2M-1)!} \cdot \frac{1}{2^{p-1}} < \frac{1}{p} \frac{(2M)^{2M}}{(2M-1)!}$$

and therefore, writing

$$H = \sum_{m=1}^{n} \{3^m a_m \sum_{r=p-1}^{q} c_r \Theta_r m^r\},$$

if

$$p > \frac{(2M)^{2M}}{(2M-1)!} \sum_{r=1}^{n} 3^{r+1} |a_r| = P, \text{ say},$$

we have $|H| < 1/3$.

Thus

$$T\varphi(0) \sum_{m=1}^{n} a_m E_N(m) = a_0 T\varphi(0) + \sum_{m=1}^{n} T\varphi(m) + H,$$

where, if p is a prime greater than $\max\{P, |a_0|\}$, we have

(i) $a_0 T\varphi(0)$ is an integer not divisible by p,

(ii) $\sum_{m=1}^{n} a_m T\varphi(m)$ is an integer divisible by p,

(iii) $|H| < 1/3$,

so that

$$|T\varphi(0) \sum_{m=0}^{n} a_m E_N(m)| > \tfrac{2}{3}.$$

Since

$$\left\{\sum_{r=0}^{n} \frac{1}{r!}\right\}^m < \sum_{r=0}^{mn} \frac{m^r}{r!}$$

therefore

$$0 < \left\{\sum_{r=0}^{n} \frac{1}{r!}\right\}^m - \left\{\sum_{r=0}^{n} \frac{m^r}{r!}\right\} < 2 \frac{m^{n+1}}{(n+1)!} < \frac{(2m)^{2m}}{(2m-1)!} \cdot \frac{1}{2^{n-1}},$$

and so

$$|T\varphi(0)[\sum_{m=0}^{n} a_m\{E_N(1)\}^m - \sum_{m=0}^{n} a_m E_N(m)]|$$

$$\leqslant 2|T\varphi(0)|\sum_{m=1}^{n}\frac{m^{N+1}}{(N+1)!}|a_m|$$

$$<|T\varphi(0)|\cdot 2An^{N+2}/(N+1)! < |T\varphi(0)|\cdot A(2n)^{2n}/(2n-1)!2^N < 1/3$$

if $N > |3AT\varphi(0)|(2n)^{2n}/(2n-1)! = B$, say.
Hence, for $N > B$ and $p > \max\{P, |a_0|\}$, we have

$$|\sum_{m=0}^{n} a_m\{E_N(1)\}^m| > 1/3|T\varphi(0)|.$$

Since $E_N(1)$ is primitive recursively convergent (with classical limit e) it follows that e is primitive recursively transcendental.

THE PRIMITIVE RECURSIVE TRANSCENDENCE OF π

We preface the proof of the primitive recursive transcendence of π by a list of the fundamental properties of the norm

$$||x+iy|| = |x|+|y|$$

of a complex number, $x+iy$, with x, y rational.
Let $x = x+iy$, $w = u+iv$, then:

1. $||z+w|| \leqslant ||z||+||w||$, for $||x+u+i(y+v)|| = |x+u|+|y+v|$
$\leqslant ||z||+||w||$.

2. $||z\pm w|| \geqslant ||z||-||w||$, for $||z-w+w|| \leqslant ||z-w||+||w||$.

3. $||zw|| \leqslant ||z||\cdot||w||$, since $||zw|| = |xu-yv|+|xv+yu|$
$\leqslant (|x|+|y|)(|u|+|v|)$.

4. $|z|^2 \leqslant ||z||^2 \leqslant 2|z|^2$.

5. $||zw|| \geqslant \frac{1}{2}||z||\cdot||w||$, for $||zw||^2 \geqslant |z|^2\cdot|w|^2 \geqslant \frac{1}{4}||z||^2\cdot||w||^2$.

6. $1/||z|| \leqslant ||1/z|| \leqslant 2/||z||$, for $||1/z|| = \{|x|+|y|\}/(x^2+y^2)$,
and $x^2+y^2 \leqslant \{|x|+|y|\}^2 \leqslant 2(x^2+y^2)$.

Let α be any algebraic number with associated polynomial

$$a_0x^N + a_1x^{N-1} + \dots + a_N$$

and let the zeros of this polynomial be

$$\alpha_1 = \alpha,\ \alpha_2,\ \alpha_3,\ \ldots,\ \alpha_N.$$

Denote $|a_0| + \ldots + |a_N|$ by $\frac{1}{2}A$ and $2\ NA$ by B.
Further let $\beta_{2r-1} = i\alpha_r$, $\beta_{2r} = -i\alpha_r$, $r = 1, 2, \ldots N$, so that for

$$1 \leqslant k \leqslant 2N,\ ||\beta_k|| < A,$$

and let γ_s, $1 \leqslant s \leqslant M = 2^{2N} - 1$, consist of the sums of the numbers β_k, $1 \leqslant k \leqslant 2N$, taken j at a time, $1 \leqslant j \leqslant 2N$, so that γ_s, $1 \leqslant s \leqslant M$, are the zeros of a polynomial

$$Q(x) = b_0 x^M + b_1 x^{M-1} + \ldots + b_M$$

with integral coefficients, and $||\gamma_s|| < B$.
We define

$$\psi(x) = \frac{x^{p-1}}{(p-1)!}\, b_0^{pM}\, \{Q(x)\}^p = c_{p-1}x^{p-1} + c_p x^p + \ldots + c_{pM+p-1}x^{pM+p-1}$$

where p is a prime greater than $|b_0\, b_M|$,

$$L(\psi(x)) = \sum_{r=1}^{\infty} \psi^r(x),\ K = L(\psi(0)) = \Sigma c_r(r!),\ E_n(x) = \sum_{r=1}^{n} x^r/(r!)$$

and

$$T_n = (1 + E_n(\beta_1))(1 + E_n(\beta_2))\ \ldots\ (1 + E_n(\beta_{2N})).$$

The numbers $\psi^r(0)$, $r \neq p-1$ are all integral multiples of p, but $\psi^{p-1}(0) = b_0^{pM}\, b_M{}^p$, an integer not divisible by p.
Furthermore if x is a zero of $Q(x)$, $\psi^r(x) = 0$ for $r \leqslant p-1$, and for $r \geqslant p$, $\sum_{m=1}^{M} \psi^r(\gamma_m)$ is an integer divisible by p.
For fixed x, $\psi(x) \to 0$ as $p \to \infty$, and si milarly

$$\sum_{r=p-1}^{pM+p-1} |c_r|\ |x|^r \to 0 \text{ as } p \to \infty.$$

Now $T_n = 1 + \sum_{r=1}^{M} E_n(\gamma_r) + V_n$, where, since

$$||E_n(x) \cdot E_n(y) - E_n(x+y)|| < 2(||x|| + ||y||)^{n+1}/(n+1)!,$$

for large enough n, and the corresponding inequality holds for three or more factors, we have

$$||V_n|| < 2MB^{n+1}/(n+1)!$$

But

$$K \cdot E_n(\gamma) = L(\psi(\gamma)) + \sum_i c_i \gamma^i R_{\gamma,i}$$

where

$$\frac{\gamma^i}{i!} R_{\gamma,i} = \frac{\gamma^{i+1}}{(i+1)!} + \frac{\gamma^{i+2}}{(i+2)!} + \ldots + \frac{\gamma^n}{n!}$$

so that $||R_{\gamma,i}|| < E_n(B)$
and therefore

$$||\sum_\gamma \sum_i c_i \gamma^i R_{\gamma,i}|| \leqslant M \cdot E_n(B) \sum |c_i| B^i \text{ tends to } 0 \text{ as } p \text{ tends to } \infty.$$

Thus

$$K \cdot \sum_\gamma E_n(\gamma) = L(\sum_\gamma \psi(\gamma)) + \varepsilon_p$$

where $\varepsilon_p \to 0$ as $p \to \infty$, and $L(\Sigma\psi(\gamma))$ is an integer divisible by p. Hence KT_n is the sum of an integer not divisible by p and a term which tends to zero with p, and a term which, for fixed p, tends to zero with n.

Choose p so that $||\varepsilon_p|| < \frac{1}{3}$, then since $2KMB^{n+1}/(n+1)! < \frac{1}{3}$ for $n \geqslant 6KMB^{B+1}/B!$, we have

$$||T_n|| > 1/3K.$$

However

$$||1 + E_n(\beta)|| < 4^A$$

and so

$$||T_n|| < ||1 + E_n(i\alpha)|| \cdot 4^{(2N-1)A} < ||1 + E_n(i\alpha)|| 4^B$$

whence

$$||1 + E_n(i\alpha)|| > 1/3K \cdot 4^B.$$

Let π_n be the value of π to n places of decimals, then it may be shown [1]) that

---

1) For details see the author's article "The recursive irrationality of π", *loc. cit.* p. 269.

(for $n \geqslant 14$)

$$||1+E_{2n+1}(i\pi_n)|| < 1/10^{n-1} < 1/12K \cdot 4^B, \text{ for } n \geqslant K \cdot 4^{B+1},$$

and so

$$||E_{2n+1}(i\alpha) - E_{2n+1}(i\pi_n)|| > 1/K \cdot 4^{B+1}, \text{ for } n \geqslant c = \max \{K \cdot 4^{B+1}, 3\,KM \cdot B^{B+1}/B!\}$$

Since $||i\pi_n|| < 4^A$ and $||i\alpha|| < 4^A$ therefore

$$||E_{2n+1}(i\alpha) - E_{2n+1}(i\pi_n)|| < ||\alpha - \pi_n|| E_{2n}(4^A) < ||\alpha - \pi_n|| \cdot 3^{(4^A)}$$

and so, for $n \geqslant c$,

$$||\alpha - \pi_n|| > 1/K \cdot 4^B \cdot 3^{(4^A)} \quad . \; . \; . \; (1)$$

whence $||\alpha - \pi|| \geqslant 1/K \cdot 4^B \cdot 3^{(4^A)}$, showing by how much π differs from α, at least.
Since (1) holds for each zero α of $a_0x^N + \ldots + a_N$ therefore, for $n \geqslant c$,

$$||a_0\pi_n{}^N + a_1\pi_n{}^{N+1} + \ldots + a_N|| \geqslant ||a_0|| \cdot ||\pi_n - \alpha_1|| \ldots ||\pi_n - \alpha_n|| \cdot 2^{-N} \geqslant ||a_0||/K^N \cdot 2^{(2^B+1)}\, 3^{N \cdot 4^A}$$

which proves that π is primitive recursively transcendental.

BIBLIOGRAPHICAL NOTES

CHAPTER I. Recursive arithmetic was introduced by Th. Skolem in Skolem [1]. Various formalisations of recursive arithmetic are given in Curry [1], Goodstein [15] and Church [2]. For an account of the properties of general recursive functions see Kleene [2] and Davis [1].

The notions of relative and recursive convergence, and the reduced sequence were introduced in Goodstein [2]. Tests for recursive convergence were first given in Goodstein [6].

For a different proof of Theorem 1, see Rice [1].

For the enumeration of primitive recursive functions see Péter [1] which also contains a very good bibliography on recursive functions.

For Specker's Theorem see Specker [1].

The proof of Theorem 1 is also due to Specker. Recursive real numbers are studied in Meschkowski [2].

CHAPTER II. The notions of recursive and relative continuity were introduced in Goodstein [2]. Theorem 2, 4 is due to Meschkowski and was published in Meschkowski [1].

Theorem 2, 5 was proved by Specker and was first published in Goodstein [5].

CHAPTER III. The notions of recursive and relative differentiability were introduced in Goodstein [2].

The classical Theorem that a function continuous in a closed interval has a maximum value in the interval is not true in recursive analysis; for proofs see Specker [2], Lacombe [1] and a review by Kreisel, in Kreisel [1] of Meschkowski [1]. Markov, in Markov [1], gives a non-constructive proof that for a recursive real function $F(x)$ such that $F(0) = -1$, $F(1) = 1$ there exists a recursive real number c, such that $F(c) = 0$; a constructive proof of this result is impossible, for Specker has shown that there is a recursive sequence F_k of recursive real functions such that $F_k(0) = -1$, $F_k(1) = 1$ and for no rational recursive function $s(k, n)$, recursively convergent in n for each k, can $\lim_{n\to\infty} F_k(s(k, n)) = 0$ hold. This shows that we cannot identify recursive analysis with a classical study of recursive real numbers. For a comparison of different levels of constructivist theories see *Constructivity in Mathematics*, edited by A. Heyting (North-Holland Publishing Co. 1959)

CHAPTER IV. The relative integral was introduced in Goodstein [11], and all the results of this chapter were first published in that paper.

Chapter V. This theory of recursive ordinals was first developed in Goodstein [1, 4].
Different constructivist treatments of ordinals are given in Ackermann [1], Hilbert–Bernays [1], Church [1], Church and Kleene [1].

Appendix. The notion of recursive irrationality (in a different terminology) was introduced in Goodstein [3], and a proof of the recursive irrationality of e^x for rational non-zero x was given in the same paper. A proof of the recursive irrationality of π is given in Goodstein [12]. Recursive transcendence was introduced in the author's paper to the Congress of Mathematicians in Edinburgh, 1958.

BIBLIOGRAPHY

ACKERMANN, W., [1] Zur Widerspruchsfreiheit der Zahlentheorie. *Math. Annalen* 117 (1940) 162–194.

[2] Konstruktiver Aufbau eines Abschnitts der zweiten Cantorschen Zahlenklasse. *Math. Zeitschrift* 53 (1951) 403–413.

BERNAYS, P., [1] Uber das Induktionsschema in der rekursiven Zahlentheorie. *Kontrolliertes Denken, Untersuchungen zum Logikkalkul und zur Logik der Einzelnwissenschaften.* Kommissions-Verlag K. Alber, München, (1951) 10–17.

CHURCH, A., [1] The constructive second number class. *Bulletin of the American Math. Soc.* 44 (1938) 224–232.

[2] Binary Recursive Arithmetic. *J. de Math.* 36 (1957), 39–55.

CHURCH, A. and S. C. KLEENE, [1] Formal definitions in the theory of ordinal numbers. *Fundamenta Mathematicae* 28 (1936) 11–21.

CURRY, H. B., [1] A formalisation of recursive arithmetic. *American Journal of Math.* 63 (1941) 263–282.

DAVIS, M., [1] *Computability and Undecidability.* New York 1958.

GOODSTEIN, R. L., [1] On the restricted ordinal theorem. *The Journal of Symbolic Logic* 9 (1944) 33–41.

[2] Function theory in an axiom-free equation calculus. *Proceedings of London Math. Soc.* (2) 48 (1945) 401–434.

[3] The strong convergence of the exponential function. *Journal of the London Math. Soc.* 22 (1947) 200–205.

[4] Transfinite ordinals in recursive number theory. *The Journal of Symbolic Logic* 12 (1947) 123–129.

[5] Mean value theorems in recursive functions theory. Part I. Differential mean value theorems. *Proceedings of the London Math. Soc.* (2) 52 (1950) 81–106

[6] The Gauss test for relative convergence. *Amer. J. Math.* 72 (1950) 217–228.

[7] *Constructive Formalism.* Essays on the foundations of mathematics. University College, Leicester. 1951.

[8] The foundations of mathematics. University College, Leicester. (1951).

[9] Permutation in recursive arithmetic. *Math. Scand.* 1 (1953) 222–226.

[10] A problem in recursive function theory. *The Journal of Symbolic Logic* 12 (1953) 225–232.

[11] The relatively exponential, logarithmic and circular functions in recursive function theory. *Acta Mathematica* 92 (1954) 171–190.

[12] The recursive irrationality of π. *The Journal of Symbolic Logic* 19 (1954) 267–274.

[13] Logic-free formalisations of recursive arithmetic. *Math. Scand.* 2 (1954) 247–267.

[14] A constructivist theory of plane curves. *Fund. Math.* 43 (1956) 23–35.

[15] *Recursive Number Theory* Amsterdam 1957.

[16] Models of propositional calculi in recursive arithmetic *Math. Scand.* 6 (1958) 293–296

[17] Recursive Analysis, in *Constructivity in Mathematics* 37–42.

Grzegorczyk, A., [1] On the definitions of computable real continuous functions. *Fund. Math.* 44 (1947) 61–71.

Hilbert, D. and P. Bernays, [1] *Grundlagen der Mathematik.* Berlin 1934, 39.

Kleene, S. C., [1] On notation for ordinal numbers. *The Journal of Symbolic Logic* 3 (1938) 150–155.

[2] *Introduction to Metamathematics.* Amsterdam 1952.

Kreisel, G., [1] Review of Meschkowski [1], *Math. Reviews* 1958, 238.

Lacombe, D., [1] Extension de la notion de fonction recursive aux fonctions d'une ou plusieurs variables réelles. *Comptes Rendus* 240, (1955) 2478–2480, 241 (1955) 13–14, 151–153.

Meschkowski, H., [1] Zur rekursiven Funktionentheorie. *Acta Math.* 95 (1956) 9–23.

[2] Rekursive reelle Zahlen. *Math. Zeitschrift* 66 (1956) 189–202.

Markov, A., [1] Uber die Stëtigkeit der konstruktiven Funktionen. *Ycnexn Mat. Hayk CCCP* 61 (1954) 226–230

Myhill, J. R., [1] Criteria of constructibility for real numbers. *The Journal of Symbolic Logic* 18 (1953) 7–10.

Peter, R., [1] *Rekursive Funktionen.* Budapest 1957.

[2] Zum Begriff der rekursiven reellen Zahl. *Acta Scient. Math.*, Szeged, 12/A. Leopoldo Fejer et Frederico Riesz LXX annos natis dedicatus. (1950) 239–245.

RICE, H. G., [1] Recursive real numbers. *Proc. Amer. Math. Soc.* 5 (1954) 784–791.

SKOLEM, TH., [1] Begründung der elementaren Arithmetik durch die rekurrierende Denkweise ohne Anwendung scheinbarer Veränderlichen mit unendlichem Ausdehnungsbereich. *Videnskapsselskapets Skrifter* 1. *Math-Naturv. Kl.* 6 (1924) 3–38.

[2] Eine Bemerkung über die Induktionsschemata in der rekursiven Zahlentheorie. *Monatshefte fur Math. und Phys.* 48 (1939) 268–276.

[3] A remark on the induction schema. *Det Kongelige Norske Videnskabers Selskab* 22 (1950) 167–170.

SPECKER, E., [1] Nicht konstruktiv beweisbare Sätze der Analysis. *The Journal of Symbolic Logic* 14 (1949) 145–158.

[2] Der Satz von Maximum in der Rekursiven Analysis. In *Constructivity in Mathematics* 254–265.

INDEX